50% OFF Online ACT Prep Course!

Dear Customer,

We consider it an honor and a privilege that you chose our ACT Study Guide. As a way of showing our appreciation and to help us better serve you, we have partnered with Mometrix Test Preparation to offer you **50% off their online ACT Course**. Many ACT courses are needlessly expensive and don't deliver enough value. With their course, you get access to the best ACT prep material, and you only pay half price.

Mometrix has structured their online course to perfectly complement your printed study guide. The ACT Online Course contains **in-depth lessons** that cover all the most important topics, **180+ video reviews** that explain difficult concepts, over **1,500 practice questions** to ensure you feel prepared, and more than **500 digital flashcards**, so you can study while you're on the go.

Online ACT Prep Course

Topics Covered:

- English
  - o Organization, Unity, and Cohesion
  - o Punctuation and Grammar
- Math
  - o Numbers and Operations
  - o Algebra and Geometry
- Reading
  - o Key Ideas and Details
  - o Purpose, Position, and Point of View
- Science
  - o Evaluation of Models
  - o Interpretation of Data
- Writing

Course Features:

- ACT Study Guide
  - o Get content that complements our best-selling study guide.
- Full-Length Practice Tests
  - o With over 1,500 practice questions, you can test yourself again and again.
- Mobile Friendly
  - o If you need to study on the go, the course is easily accessible from your mobile device.
- ACT Flashcards
  - o Their course includes a flashcards mode with over 500 content cards for you to study.

To receive this discount, visit their website: mometrix.com/university/act and add the course to your cart. At the checkout page, enter the discount code: **APEXACT50**

If you have any questions or concerns, please contact them at universityhelp@mometrix.com.

Sincerely,

 in partnership with

FREE

Free Study Tips DVD

In addition to the tips and content in this guide, we have created a FREE DVD with helpful study tips to further assist your exam preparation. **This FREE Study Tips DVD provides you with top-notch tips to conquer your exam and reach your goals.**

Our simple request in exchange for the strategy-packed DVD is that you email us your feedback about our study guide. We would love to hear what you thought about the guide, and we welcome any and all feedback—positive, negative, or neutral. It is our #1 goal to provide you with top quality products and customer service.

To receive your **FREE Study Tips DVD**, email freedvd@apexprep.com. Please put "FREE DVD" in the subject line and put the following in the email:

a. The name of the study guide you purchased.

b. Your rating of the study guide on a scale of 1-5, with 5 being the highest score.

c. Any thoughts or feedback about your study guide.

d. Your first and last name and your mailing address, so we know where to send your free DVD!

Thank you!

ACT Prep Book 2021 and 2022

ACT Study Guide with Practice Test Questions for All Sections [English, Math, Reading, Science, Essay]

Matthew Lanni

Written and edited by APEX Publishing.

ISBN 13: 9781628457599
ISBN 10: 1628457597

APEX Publishing is not connected with or endorsed by any official testing organization. APEX Publishing creates and publishes unofficial educational products. All test and organization names are trademarks of their respective owners.

The material in this publication is included for utilitarian purposes only and does not constitute an endorsement by APEX Publishing of any particular point of view.

For additional information or for bulk orders, contact info@apexprep.com.

Table of Contents

Test Taking Strategies

1. Reading the Whole Question

A popular assumption in Western culture is the idea that we don't have enough time for anything. We speed while driving to work, we want to read an assignment for class as quickly as possible, or we want the line in the supermarket to dwindle faster. However, speeding through such events robs us from being able to thoroughly appreciate and understand what's happening around us. While taking a timed test, the feeling one might have while reading a question is to find the correct answer as quickly as possible. Although pace is important, don't let it deter you from reading the whole question. Test writers know how to subtly change a test question toward the end in various ways, such as adding a negative or changing focus. If the question has a passage, carefully read the whole passage as well before moving on to the questions. This will help you process the information in the passage rather than worrying about the questions you've just read and where to find them. A thorough understanding of the passage or question is an important way for test takers to be able to succeed on an exam.

2. Examining Every Answer Choice

Let's say we're at the market buying apples. The first apple we see on top of the heap may *look* like the best apple, but if we turn it over we can see bruising on the skin. We must examine several apples before deciding which apple is the best. Finding the correct answer choice is like finding the best apple. Although it's tempting to choose an answer that seems correct at first without reading the others, it's important to read each answer choice thoroughly before making a final decision on the answer. The aim of a test writer might be to get as close as possible to the correct answer, so watch out for subtle words that may indicate an answer is incorrect. Once the correct answer choice is selected, read the question again and the answer in response to make sure all your bases are covered.

3. Eliminating Wrong Answer Choices

Sometimes we become paralyzed when we are confronted with too many choices. Which frozen yogurt flavor is the tastiest? Which pair of shoes look the best with this outfit? What type of car will fill my needs as a consumer? If you are unsure of which answer would be the best to choose, it may help to use process of elimination. We use "filtering" all the time on sites such as eBay® or Craigslist® to eliminate the ads that are not right for us. We can do the same thing on an exam. Process of elimination is crossing out the answer choices we know for sure are wrong and leaving the ones that might be correct. It may help to cover up the incorrect answer choice. Covering incorrect choices is a psychological act that alleviates stress due to the brain being exposed to a smaller amount of information. Choosing between two answer choices is much easier than choosing between all of them, and you have a better chance of selecting the correct answer if you have less to focus on.

4. Sticking to the World of the Question

When we are attempting to answer questions, our minds will often wander away from the question and what it is asking. We begin to see answer choices that are true in the real world instead of true in the world of the question. It may be helpful to think of each test question as its own little world. This world may be different from ours. This world may know as a truth that the chicken came before the egg or may assert that two plus two equals five. Remember that, no matter what hypothetical nonsense may be in the question, assume it to be true. If the question states that the chicken came before the egg, then choose your answer based on that truth. Sticking to the world of the question means placing all of our biases and assumptions aside and relying on the question to guide us to the correct answer. If we are simply looking for answers that are correct based on our own judgment, then we may choose incorrectly. Remember an answer that is true does not necessarily answer the question.

5. Key Words

If you come across a complex test question that you have to read over and over again, try pulling out some key words from the question in order to understand what exactly it is asking. Key words may be words that surround the question, such as *main idea, analogous, parallel, resembles, structured,* or *defines.* The question may be asking for the main idea, or it may be asking you to define something. Deconstructing the sentence may also be helpful in making the question simpler before trying to answer it. This means taking the sentence apart and obtaining meaning in pieces, or separating the question from the foundation of the question. For example, let's look at this question:

> Given the author's description of the content of paleontology in the first paragraph, which of the following is most parallel to what it taught?

The question asks which one of the answers most *parallels* the following information: The *description* of paleontology in the first paragraph. The first step would be to see *how* paleontology is described in the first paragraph. Then, we would find an answer choice that parallels that description. The question seems complex at first, but after we deconstruct it, the answer becomes much more attainable.

6. Subtle Negatives

Negative words in question stems will be words such as *not, but, neither,* or *except.* Test writers often use these words in order to trick unsuspecting test takers into selecting the wrong answer—or, at least, to test their reading comprehension of the question. Many exams will feature the negative words in all caps (*which of the following is NOT an example*), but some questions will add the negative word seamlessly into the sentence. The following is an example of a subtle negative used in a question stem:

> According to the passage, which of the following is *not* considered to be an example of paleontology?

If we rush through the exam, we might skip that tiny word, *not,* inside the question, and choose an answer that is opposite of the correct choice. Again, it's important to read the question fully, and double check for any words that may negate the statement in any way.

7. Spotting the Hedges

The word "hedging" refers to language that remains vague or avoids absolute terminology. Absolute terminology consists of words like *always, never, all, every, just, only, none,* and *must.* Hedging refers to words like *seem, tend, might, most, some, sometimes, perhaps, possibly, probability,* and *often.* In some cases, we want to choose answer choices that use hedging and avoid answer choices that use absolute terminology. It's important to pay attention to what subject you are on and adjust your response accordingly.

8. Restating to Understand

Every now and then we come across questions that we don't understand. The language may be too complex, or the question is structured in a way that is meant to confuse the test taker. When you come across a question like this, it may be worth your time to rewrite or restate the question in your own words in order to understand it better. For example, let's look at the following complicated question:

> Which of the following words, if substituted for the word *parochial* in the first paragraph, would LEAST change the meaning of the sentence?

Let's restate the question in order to understand it better. We know that they want the word *parochial* replaced. We also know that this new word would "least" or "not" change the meaning of the sentence. Now let's try the sentence again:

> Which word could we replace with *parochial,* and it would not change the meaning?

Restating it this way, we see that the question is asking for a synonym. Now, let's restate the question so we can answer it better:

> Which word is a synonym for the word *parochial?*

Before we even look at the answer choices, we have a simpler, restated version of a complicated question.

9. Predicting the Answer

After you read the question, try predicting the answer *before* reading the answer choices. By formulating an answer in your mind, you will be less likely to be distracted by any wrong answer choices. Using predictions will also help you feel more confident in the answer choice you select. Once you've chosen your answer, go back and reread the question and answer choices to make sure you have the best fit. If you have no idea what the answer may be for a particular question, forego using this strategy.

10. Avoiding Patterns

One popular myth in grade school relating to standardized testing is that test writers will often put multiple-choice answers in patterns. A runoff example of this kind of thinking is that the most common answer choice is "C," with "B" following close behind. Or, some will advocate certain made-up word patterns that simply do not exist. Test writers do not arrange their correct answer choices in any kind of pattern; their choices are randomized. There may even be times where the correct answer choice will be the same letter for two or three questions in a row, but we have no way of knowing when or if this might happen. Instead of trying to figure out what choice the test writer probably set as being correct, focus on what the *best answer choice* would be out of the answers you are presented with. Use the tips above, general knowledge, and reading comprehension skills in order to best answer the question, rather than looking for patterns that do not exist.

FREE DVD OFFER

Achieving a high score on your exam depends not only on understanding the content, but also on understanding how to apply your knowledge and your command of test taking strategies. **Because your success is our primary goal, we offer a FREE Study Tips DVD, which provides top-notch test taking strategies to help you optimize your testing experience.**

Our simple request in exchange for the strategy-packed DVD is that you email us your feedback about our study guide.

To receive your **FREE Study Tips DVD**, email freedvd@apexprep.com. Please put "FREE DVD" in the subject line and put the following in the email:

a. The name of the study guide you purchased.

b. Your rating of the study guide on a scale of 1-5, with 5 being the highest score.

c. Any thoughts or feedback about your study guide.

d. Your first and last name and your mailing address, so we know where to send your free DVD!

Introduction to the ACT

Function of the Test

The ACT is one of the two national standardized college entrance examinations, with the SAT serving as the other option. Most hopeful college-bound students take the ACT, SAT, or both. For admissions purposes, every four-year college and university in the United States accepts ACE scores, and some schools require it. More so than the SAT, which primarily serves as an aptitude test, the ACT is often used for course placement purposes because it measures academic achievement on content addressed in high school classes. Twelve states also require that all high school juniors in the state take the ACT, and eight additional states have counties that require the exam.

Most ACT test takers are prospective college students who are currently in their junior or senior year of high school. More than 2 millions students graduating in the class of 2017 took the ACT.

Test Administration

The ACT is offered on seven dates in the U.S. and Canada, and on six dates internationally throughout the year. The exam is usually administered at high schools or colleges, but other locations may be offered. The registration fee includes the cost to submit score reports to four colleges, but for an additional fee, students can send scores to additional institutions. There is a separate registration fee incurred for the optional writing section. Some high schools cover the fees for their students, so prospective test takers are advised to contact the guidance counselor at their school.

Test takers can retake the ACT every time the test is offered, up to a maximum of twelve times. However, different colleges and universities sometimes have limits on the number of retakes they will consider. Beginning September 2020, test takers wishing to take one or several of the five sections of the ACT can do so without needing to retake all sections. This can be advantageous for test takers who do not want to risk lowering their scores on sections over which they previously performed well.

Reasonable accommodations will be provided to test takers with appropriate documentation for a variety of disabilities.

Test Format

Test takers are given a total of 175 minutes to complete the 215 multiple-choice questions in four subject subtests (English, Mathematics, Reading, and Science) of the ACT. It also has an optional Writing Test, which involves writing an essay, which takes an additional forty minutes. Some colleges and universities require the essay for admission.

The English Test consists of 75 questions that address the production of writing, knowledge of language, and conventions of standard English. The 60-question Mathematics Test involves number sense, algebra, functions, geometry, statistics and probability, and modeling. Calculators that meet certain calculator requirements are permitted. The Reading Test is contains four written passages, with ten questions per passage addressing comprehension skills, the ability to make inferences and draw conclusions, and apply and integrate knowledge. The Science Test contains 40 questions that require interpreting data, understanding scientific investigations, and evaluating models and results.

The Writing Test is always given at the end of the exam so that test takers opting not to take it may leave after completing the other four subtests. This section consists of one essay in which students must analyze three different perspectives on a broad social issue and reconcile them in a cohesive essay. The following chart provides the breakdown of the sections of the ACT:

Subtest	Length	Number of Questions
English	45 minutes	75
Mathematics	60 minutes	60
Reading	35 minutes	40
Science	35 minutes	40
Writing (optional)	40 minutes	1 essay

Scoring

Score reports are typically available two weeks after the date of administration. Because there is no penalty for incorrect answers, test takers are encouraged to answer every question, even if they have to guess. For each of the four required subtests, test takers receive a score between 1 and 36. These scores are then averaged together to yield a Composite Score, which is the primary score reported as an "ACT score." The most prestigious colleges and universities are typically looking for Composite Scores greater than 30 in order to consider an applicant for admissions. Other selective schools typically expect candidates to have scores just under 30. Average institutions are more likely to set the bar lower (perhaps in the low 20s), while community colleges usually accept students with scores in the high teens. In 2016 and 2017, the mean Composite Score among all test takers (including those not applying to college) was 20.9.

The Writing Test is scored on a scale that ranges from 2 to 12 scale. In 2016 and 2017, the mean score was 6.7.

Prior to September 2020, test takers who chose to retake the exam were required to retake all sections of the exam. Then, their composite score would simply be reflective of their performance on the subsections of the retake. Scores from individual subsections from different test attempts could not be combined. As of September 2020, a new "superscore" can be generated. This score is a conglomerate of the test taker's highest scores on subsections from any of their attempts at the test.

English Test

Production of Writing

Topic Development

Rhetorical Aspects of Texts

An essay presents an idea to the reader from its author's perspective. The central assertion is stated in the first sentence of the essay, the thesis sentence. It is followed by paragraphs containing supporting ideas, which are developed by means of facts, details, and examples. The essay ends with a conclusion that paraphrases the thesis sentence, sums up the major points, and gives the reader a feeling of closure.

Identifying the Purpose of Different Parts of a Text

An essay contains a beginning, a middle, and an end. The purpose of the beginning, or introduction, is for the author to state the idea that she wants to explain in the middle, or body, of the essay. The body is where the author builds her argument with facts and examples that explain, support, and show the application of the thesis statement in the introduction, and by analyzing and interpreting the facts for the reader. In the end, or conclusion, the author restates the thesis and then brings the essay to a close by establishing the significance of the thesis in a larger context, or by posing a question that will compel further thought on the part of the reader.

Determining Whether a Text or Part of a Text has Met its Goal

An essay, or part of an essay, meets its goal by communicating an idea clearly and effectively to the reader. Clear communication is achieved through logical organization of the information that supports each main point. Arguments that are well thought out and communicated in a suitable tone are the foundations of effective communication. An essay has unity if all the parts relate to the thesis sentence, and it has coherence if the reader can follow the author's thoughts as they progress. If the reader is motivated, the essay has succeeded.

Evaluating the Relevance of Material in Terms of a Text's Focus

An essay should have its own internal logic, and each idea and element in it should fit together. If something doesn't fit, chances are it is irrelevant information that will distract the reader.

Organization, Unity, and Cohesion

Developing a Well-Organized Paragraph

A **paragraph** is a series of connected and related sentences addressing one topic. Writing good paragraphs benefits writers by helping them to stay on target while drafting and revising their work. It benefits readers by helping them to follow the writing more easily. Regardless of how brilliant their ideas may be, writers who do not present these ideas in organized ways will fail to engage readers—and fail to accomplish their writing goals. A fundamental rule for paragraphing is to confine each paragraph to a single idea. When writers find themselves transitioning to a new idea, they should start a new paragraph. However, a paragraph can include several pieces of evidence supporting its single idea, and it can include several points if they are all related to the overall paragraph topic. When writers find each point becoming lengthy, they may choose instead to devote a separate paragraph to every point and elaborate upon each more fully.

An effective paragraph should have the following elements:

- **Unity:** One major discussion point or focus should occupy the whole paragraph from beginning to end.

- **Coherence:** For readers to understand a paragraph, it must be coherent. Two components of coherence are logical and verbal bridges. In logical bridges, the writer may write consecutive sentences with parallel structure or carry an idea over across sentences. In verbal bridges, writers may repeat key words across sentences.

- A **topic sentence:** The paragraph should have a sentence that generally identifies the paragraph's thesis or main idea.

- Sufficient **development:** To develop a paragraph, writers can use the following techniques after stating their topic sentence:

  - Define terms

  - Cite data

  - Use illustrations, anecdotes, and examples

  - Evaluate causes and effects

  - Analyze the topic

  - Explain the topic using chronological order

A topic sentence identifies the main idea of the paragraph. Some are explicit, while some are implicit. The topic sentence can appear anywhere in the paragraph. However, many experts advise beginning writers to place each paragraph topic sentence at or near the beginning of its paragraph to ensure that their readers understand what the topic of each paragraph is. Even without having written an explicit topic sentence, the writer should still be able to summarize readily what subject matter each paragraph addresses. The writer must then fully develop the topic that is introduced or identified in the topic sentence. Depending on what the writer's purpose is, they may use different methods for developing each paragraph.

Two main steps in the process of organizing paragraphs and essays should both be completed after determining the writing's main point, while the writer is planning or outlining the work. The initial step is to give an order to the topics addressed in each paragraph. Writers must have logical reasons for putting one paragraph first, another second, etc. The second step is to sequence the sentences in each paragraph. As with the first step, writers must have logical reasons for the order of sentences. Sometimes the work's main point obviously indicates a specific order.

Topic Sentences
To be effective, a topic sentence should be concise so that readers get its point without losing the meaning among too many words. As an example, in *Only Yesterday: An Informal History of the 1920s* (1931), author Frederick Lewis Allen's topic sentence introduces his paragraph describing the 1929 stock market crash: "The Bull Market was dead." This example illustrates the criteria of conciseness and brevity. It is also a strong sentence, expressed clearly and unambiguously. The topic sentence also introduces the paragraph, alerting the reader's attention to the main idea of the paragraph and the subject matter that follows the topic sentence.

Experts often recommend opening a paragraph with the topic sentence to enable the reader to realize the main point of the paragraph immediately. Application letters for jobs and university admissions also benefit from opening with topic sentences. However, positioning the topic sentence at the end of a paragraph is more logical when the paragraph identifies a number of specific details that accumulate evidence and then culminates with a generalization. While paragraphs with extremely obvious main ideas need no topic sentences, more often—and particularly for students learning to write—the topic sentence is the most important sentence in the paragraph. It not only communicates the main idea quickly to readers, but it also helps writers produce and control information.

Knowledge of Language

Context Clues

Readers can often figure out what unfamiliar words mean without interrupting their reading to look them up in dictionaries by examining context. **Context** includes the other words or sentences in a passage. One common context clue is the root word and any affixes (prefixes/suffixes). Another common context clue is a synonym or definition included in the sentence. Sometimes both exist in the same sentence. Here's an example:

> Scientists who study birds are *ornithologists*.

Many readers may not know the word *ornithologist*. However, the example contains a definition (scientists who study birds). The reader may also have the ability to analyze the suffix (*-logy*, meaning the study of) and root (*ornitho-*, meaning bird).

Another common context clue is a sentence that shows differences. Here's an example:

> Birds *incubate* their eggs outside of their bodies, unlike mammals.

Some readers may be unfamiliar with the word *incubate*. However, since we know that "unlike mammals," birds incubate their eggs outside of their bodies, we can infer that *incubate* has something to do with keeping eggs warm outside the body until they are hatched.

In addition to analyzing the etymology of a word's root and affixes and extrapolating word meaning from sentences that contrast an unknown word with an antonym, readers can also determine word meanings from sentence context clues based on logic. Here's an example:

> Birds are always looking out for predators that could attack their young.

The reader who is unfamiliar with the word *predator* could determine from the context of the sentence that predators usually prey upon baby birds and possibly other young animals. Readers might also use the context clue of etymology here, as *predator* and *prey* have the same root.

Analyzing Word Parts

By learning some of the etymologies of words and their parts, readers can break new words down into components and analyze their combined meanings. For example, the root word *soph* is Greek for wise or knowledge. Knowing this informs the meanings of English words including *sophomore, sophisticated,* and *philosophy*. Those who also know that *phil* is Greek for love will realize that *philosophy* means the love of knowledge. They can then extend this knowledge of *phil* to understand *philanthropist* (one who loves people), *bibliophile* (book lover), *philharmonic* (loving harmony), *hydrophilic* (water-loving), and so on. In addition, *phob-* derives from the Greek *phobos,* meaning fear. This informs all words ending with it as meaning fear of various things: *acrophobia* (fear of heights), *arachnophobia* (fear of spiders), *claustrophobia* (fear of enclosed spaces), *ergophobia* (fear of work), and *hydrophobia* (fear of water), among others.

Some English word origins from other languages, like ancient Greek, are found in large numbers and varieties of English words. An advantage of the shared ancestry of these words is that once readers recognize the meanings of some Greek words or word roots, they can determine or at least get an idea of what many different English words mean. As an example, the Greek word *métron* means to measure, a measure, or something used to measure; the English word meter derives from it. Knowing this informs many other English words, including *altimeter, barometer, diameter, hexameter, isometric,* and *metric.* While readers must know the meanings of the other parts of these words to decipher their meaning fully, they already have an idea that they are all related in some way to measures or measuring.

While all English words ultimately derive from a proto-language known as Indo-European, many of them historically came into the developing English vocabulary later, from sources like the ancient Greeks' language, the Latin used throughout Europe and much of the Middle East during the reign of the Roman Empire, and the Anglo-Saxon languages used by England's early tribes. In addition to classic revivals and native foundations, by the Renaissance era other influences included French, German, Italian, and Spanish. Today we can often discern English word meanings by knowing common roots and affixes, particularly from Greek and Latin.

The following is a list of common prefixes and their meanings:

Prefix	Definition	Examples
a-	without	atheist, agnostic
ad-	to, toward	advance
ante-	before	antecedent, antedate
anti-	opposing	antipathy, antidote
auto-	self	autonomy, autobiography
bene-	well, good	benefit, benefactor
bi-	two	bisect, biennial
bio-	life	biology, biosphere
chron-	time	chronometer, synchronize
circum-	around	circumspect, circumference
com-	with, together	commotion, complicate
contra-	against, opposing	contradict, contravene
cred-	belief, trust	credible, credit
de-	from	depart
dem-	people	demographics, democracy
dis-	away, off, down, not	dissent, disappear
equi-	equal, equally	equivalent
ex-	former, out of	extract
for-	away, off, from	forget, forswear
fore-	before, previous	foretell, forefathers
homo-	same, equal	homogenized
hyper-	excessive, over	hypercritical, hypertension

Prefix	Definition	Examples
in-	in, into	intrude, invade
inter-	among, between	intercede, interrupt
mal-	bad, poorly, not	malfunction
micr-	small	microbe, microscope
mis-	bad, poorly, not	misspell, misfire
mono-	one, single	monogamy, monologue
mor-	die, death	mortality, mortuary
neo-	new	neolithic, neoconservative
non-	not	nonentity, nonsense
omni-	all, everywhere	omniscient
over-	above	overbearing
pan-	all, entire	panorama, pandemonium
para-	beside, beyond	parallel, paradox
phil-	love, affection	philosophy, philanthropic
poly-	many	polymorphous, polygamous
pre-	before, previous	prevent, preclude
prim-	first, early	primitive, primary
pro-	forward, in place of	propel, pronoun
re-	back, backward, again	revoke, recur
sub-	under, beneath	subjugate, substitute
super-	above, extra	supersede, supernumerary
trans-	across, beyond, over	transact, transport
ultra-	beyond, excessively	ultramodern, ultrasonic, ultraviolet
un-	not, reverse of	unhappy, unlock
vis-	to see	visage, visible

The following is a list of common suffixes and their meanings:

Suffix	Definition	Examples
-able	likely, able to	capable, tolerable
-ance	act, condition	acceptance, vigilance
-ard	one that does excessively	drunkard, wizard
-ation	action, state	occupation, starvation
-cy	state, condition	accuracy, captaincy
-er	one who does	teacher
-esce	become, grow, continue	convalesce, acquiesce
-esque	in the style of, like	picturesque, grotesque
-ess	feminine	waitress, lioness
-ful	full of, marked by	thankful, zestful
-ible	able, fit	edible, possible, divisible
-ion	action, result, state	union, fusion
-ish	suggesting, like	churlish, childish
-ism	act, manner, doctrine	barbarism, socialism
-ist	doer, believer	monopolist, socialist

Suffix	Definition	Examples
-ition	action, result, state,	sedition, expedition
-ity	quality, condition	acidity, civility
-ize	cause to be, treat with	sterilize, mechanize, criticize
-less	lacking, without	hopeless, countless
-like	like, similar	childlike, dreamlike
-ly	like, of the nature of	friendly, positively
-ment	means, result, action	refreshment, disappointment
-ness	quality, state	greatness, tallness
-or	doer, office, action	juror, elevator, honor
-ous	marked by, given to	religious, riotous
-some	apt to, showing	tiresome, lonesome
-th	act, state, quality	warmth, width
-ty	quality, state	enmity, activity

Conventions of Standard English Spelling

Homophones

Homophones are words that have different meanings and spellings, but sound the same. These can be confusing for English Language Learners (ELLs) and beginning students, but even native English-speaking adults can find them problematic unless informed by context. Whereas listeners must rely entirely on context to differentiate spoken homophone meanings, readers with good spelling knowledge have a distinct advantage since homophones are spelled differently. For instance, *their* means belonging to them; *there* indicates location; and *they're* is a contraction of *they are*; despite different meanings, they all sound the same. *Lacks* can be a plural noun or a present-tense, third-person singular verb; either way it refers to absence—*deficiencies* as a plural noun, and *is deficient in* as a verb. But *lax* is an adjective that means loose, slack, relaxed, uncontrolled, or negligent. These two spellings, derivations, and meanings are completely different. With speech, listeners cannot know spelling and must use context; however, with print, readers with spelling knowledge can differentiate them with or without context.

Homonyms, Homophones, and Homographs

Homophones are words that sound the same in speech, but have different spellings and meanings. For example, *to, too,* and *two* all sound alike, but have three different spellings and meanings. Homophones with different spellings are also called **heterographs.** **Homographs** are words that are spelled identically, but have different meanings. If they also have different pronunciations, they are **heteronyms.** For instance, *tear* pronounced one way means a drop of liquid formed by the eye; pronounced another way, it means to rip. Homophones that are also homographs are **homonyms.** For example, *bark* can mean the outside of a tree or a dog's vocalization; both meanings have the same spelling. *Stalk* can mean a plant stem or to pursue and/or harass somebody; these are spelled and pronounced the same. *Rose* can mean a flower or the past tense of *rise*. Many non-linguists confuse things by using "homonym" to mean sets of words that are homophones but not homographs, and also those that are homographs but not homophones.

The word *row* can mean to use oars to propel a boat; a linear arrangement of objects or print; or an argument. It is pronounced the same with the first two meanings, but differently with the third. Because it is spelled identically regardless, all three meanings are homographs. However, the two meanings pronounced the same are homophones, whereas the one with the different pronunciation is a heteronym. By contrast, the word *read* means to peruse language, whereas the word *reed* refers to a marsh plant.

Because these are pronounced the same way, they are homophones; because they are spelled differently, they are heterographs. Homonyms are both homophones and homographs—pronounced and spelled identically, but with different meanings. One distinction between homonyms is of those with separate, unrelated etymologies, called "true" homonyms, e.g. *skate* meaning a fish or *skate* meaning to glide over ice/water. Those with common origins are called polysemes or polysemous homonyms, e.g. the *mouth* of an animal/human or of a river.

Irregular Plurals
One type of irregular English plural involves words that are spelled the same whether they are singular or plural. These include *deer, fish, salmon, trout, sheep, moose, offspring, species, aircraft,* etc. The spelling rule for making these words plural is simple: they do not change. Another type of irregular English plurals does change from singular to plural form, but it does not take regular English *–s* or *–es* endings. Their irregular plural endings are largely derived from grammatical and spelling conventions in the other languages of their origins, like Latin, German, and vowel shifts and other linguistic mutations. Some examples of these words and their irregular plurals include *child* and *children; die* and *dice; foot* and *feet; goose* and *geese; louse* and *lice; man* and *men; mouse* and *mice; ox* and *oxen; person* and *people; tooth* and *teeth;* and *woman* and *women.*

Contractions
Contractions are formed by joining two words together, omitting one or more letters from one of the component words, and replacing the omitted words with an apostrophe. An obvious yet often forgotten rule for spelling contractions is to place the apostrophe where the letters were omitted; for example, spelling errors like *did'nt* for *didn't. Didn't* is a contraction of *did not.* Therefore, the apostrophe replaces the "o" that is omitted from the "not" component. Another common error is confusing contractions with possessives because both include apostrophes, e.g. spelling the possessive *its* as "it's," which is a contraction of "it is"; spelling the possessive *their* as "they're," a contraction of "they are"; spelling the possessive *whose* as "who's," a contraction of "who is"; or spelling the possessive *your* as "you're," a contraction of "you are."

Frequently Misspelled Words
One source of spelling errors is not knowing whether to drop the final letter *e* from a word when its form is changed by adding an ending to indicate the past tense or progressive participle of a verb, converting an adjective to an adverb, a noun to an adjective, etc. Some words retain the final *e* when another syllable is added; others lose it. For example, *true* becomes *truly; argue* becomes *arguing; come* becomes *coming; write* becomes *writing;* and *judge* becomes *judging.* In these examples, the final *e* is dropped before adding the ending. But *severe* becomes *severely; complete* becomes *completely; sincere* becomes *sincerely; argue* becomes *argued;* and *care* becomes *careful.* In these instances, the final *e* is retained before adding the ending. Note that some words, like *argue* in these examples, drops the final *e* when the *–ing* ending is added to indicate the participial form; but the regular past tense ending of *–ed* makes it *argued,* in effect replacing the final *e* so that *arguing* is spelled without an *e* but *argued* is spelled with one.

Some English words contain the vowel combination of *ei,* while some contain the reverse combination of *ie.* Many people confuse these. Some examples include these:

ceiling, conceive, leisure, receive, weird, their, either, foreign, sovereign, neither, neighbors, seize, forfeit, counterfeit, height, weight, protein, and *freight*

Words with *ie* include *piece, believe, chief, field, friend, grief, relief, mischief, siege, niece, priest, fierce, pierce, achieve, retrieve, hygiene, science,* and *diesel.* A rule that also functions as a mnemonic device is "I before E except after C, or when sounded like A as in 'neighbor' or 'weigh'." However, it is obvious from the list above that many exceptions exist.

Many people often misspell certain words by confusing whether they have the vowel *a, e,* or *i,* frequently in the middle syllable of three-syllable words or beginning the last syllables that sound the same in different words. For example, in the following correctly spelled words, the vowel in boldface is the one people typically get wrong by substituting one or either of the others for it:

> cem**e**tery, quant**i**ties, ben**e**fit, privil**e**ge, unpleas**a**nt, sep**a**rate, independ**e**nt, excell**e**nt, cat**e**gories, indispens**a**ble, and irrelev**a**nt

The words with final syllables that sound the same when spoken but are spelled differently include *unpleasant, independent, excellent,* and *irrelevant.* Another source of misspelling is whether or not to double consonants when adding suffixes. For example, we double the last consonant before *–ed* and *–ing* endings in *controlled, beginning, forgetting, admitted, occurred, referred,* and *hopping;* but we do not double the last consonant before the suffix in *shining, poured, sweating, loving, hating, smiling,* and *hoping.*

One way in which people misspell certain words frequently is by failing to include letters that are silent. Some letters are articulated when pronounced correctly but elided in some people's speech, which then transfers to their writing. Another source of misspelling is the converse: people add extraneous letters. For example, some people omit the silent *u* in **guarantee,** overlook the first *r* in *surprise,* leave out the *z* in *realize,* fail to double the *m* in *recommend,* leave out the middle *i* from *aspirin,* and exclude the *p* from *temperature.* The converse error, adding extra letters, is common in words like *until* by adding a second *l* at the end; or by inserting a superfluous syllabic *a* or *e* in the middle of *athletic,* reproducing a common mispronunciation.

Consistency in Style and Tone

To improve coherence and flow in a piece of writing, it is important for a writer to remain consistent in the style and tone used. The piece of writing should feel unified, and as if its author has maintained the same "voice" throughout. One way to ensure consistency in style and tone is maintained is to read the completed writing in its entirety in one sitting, "listening" for any deviations in word choice or structure that noticeably change the tone or style of the writing.

Conventions of Standard English

Sentence Structure and Formation

Four types of improper sentences are sentence fragments, run-on sentences, subject-verb and/or pronoun-antecedent disagreement, and non-parallel structure.

Sentence Fragments
Sentence fragments are caused by absent subjects, absent verbs, or dangling/uncompleted dependent clauses. Every sentence must have a subject and a verb to be complete. An example of a **fragment** is "Raining all night long," because there is no subject present. "It was raining all night long" is one correction. Another example of a sentence fragment is the second part in "Many scientists think in unusual ways. Einstein, for instance." The second phrase is a fragment because it has no verb. One correction is "Many scientists, like Einstein, think in unusual ways." Finally, look for "cliffhanger" words like *if, when, because,* or *although* that introduce **dependent clauses**, which cannot stand alone without an

independent clause. For example, to correct the sentence fragment "If you get home early," add an independent clause: "If you get home early, we can go dancing."

Run-On Sentences

A run-on sentence combines two or more complete sentences without punctuating them correctly or separating them. For example, a run-on sentence caused by a lack of punctuation is the following:

> There is a malfunction in the computer system however there is nobody available right now who knows how to troubleshoot it.

One correction is, "There is a malfunction in the computer system; however, there is nobody available right now who knows how to troubleshoot it." Another is, "There is a malfunction in the computer system. However, there is nobody available right now who knows how to troubleshoot it."

An example of a **comma splice** of two sentences is the following:

> Jim decided not to take the bus, he walked home.

Replacing the comma with a period or a semicolon corrects this. Commas that try and separate two independent clauses without a contraction are considered comma splices.

Parallel Sentence Structures

Parallel structure in a sentence matches the forms of sentence components. Any sentence containing more than one description or phrase should keep them consistent in wording and form. Readers can easily follow writers' ideas when they are written in parallel structure, making it an important element of correct sentence construction. For example, this sentence lacks parallelism: "Our coach is a skilled manager, a clever strategist, and works hard." The first two phrases are parallel, but the third is not. Correction: "Our coach is a skilled manager, a clever strategist, and a hard worker." Now all three phrases match in form. Here is another example:

> Fred intercepted the ball, escaped tacklers, and a touchdown was scored.

This is also non-parallel. Here is the sentence corrected:

> Fred intercepted the ball, escaped tacklers, and scored a touchdown.

Sentence Fluency

For fluent composition, writers must use a variety of sentence types and structures, and also ensure that they smoothly flow together when they are read. To accomplish this, they must first be able to identify fluent writing when they read it. This includes being able to distinguish among simple, compound, complex, and compound-complex sentences in text; to observe variations among sentence types, lengths, and beginnings; and to notice figurative language and understand how it augments sentence length and imparts musicality. Once writers recognize superior fluency, they should revise their own writing to be more readable and fluent. They must be able to apply acquired skills to revisions before being able to apply them to new drafts.

One strategy for revising writing to increase its sentence fluency is flipping sentences. This involves rearranging the word order in a sentence without deleting, changing, or adding any words. For example, the student or other writer who has written the sentence, "We went bicycling on Saturday" can revise it to, "On Saturday, we went bicycling." Another technique is using appositives. An **appositive** is a phrase or word that renames or identifies another adjacent word or phrase. Writers can revise for sentence fluency by inserting main phrases/words from one shorter sentence into another shorter sentence, combining them into one longer sentence, e.g. from "My cat Peanut is a gray and brown tabby. He loves hunting rats." to "My cat Peanut, a gray and brown tabby, loves hunting rats." Revisions can also connect shorter sentences by using conjunctions and commas and removing repeated words: "Scott likes eggs. Scott is allergic to eggs" becomes "Scott likes eggs, but he is allergic to them."

One technique for revising writing to increase sentence fluency is "padding" short, simple sentences by adding phrases that provide more details specifying why, how, when, and/or where something took place. For example, a writer might have these two simple sentences: "I went to the market. I purchased a cake." To revise these, the writer can add the following informative dependent and independent clauses and prepositional phrases, respectively: "Before my mother woke up, I sneaked out of the house and went to the supermarket. As a birthday surprise, I purchased a cake for her." When revising sentences to make them longer, writers must also punctuate them correctly to change them from simple sentences to compound, complex, or compound-complex sentences.

Skills Writers Can Employ to Increase Fluency
One way writers can increase fluency is by varying the beginnings of sentences. Writers do this by starting most of their sentences with different words and phrases rather than monotonously repeating the same ones across multiple sentences. Another way writers can increase fluency is by varying the lengths of sentences. Since run-on sentences are incorrect, writers make sentences longer by also converting them from simple to compound, complex, and compound-complex sentences. The coordination and subordination involved in these also give the text more variation and interest, hence more fluency. Here are a few more ways writers can increase fluency:

- Varying the transitional language and conjunctions used makes sentences more fluent.
- Writing sentences with a variety of rhythms by using prepositional phrases.
- Varying sentence structure adds fluency.

Punctuation

Rules of Capitalization
The first word of any document, and of each new sentence, is capitalized. Proper nouns, like names and adjectives derived from proper nouns, should also be capitalized. Here are some examples:

- Grand Canyon
- Pacific Palisades
- Golden Gate Bridge
- Freudian slip
- Shakespearian, Spenserian, or Petrarchan sonnet
- Irish song

Some exceptions are adjectives, originally derived from proper nouns, which through time and usage are no longer capitalized, like *quixotic, herculean*, or *draconian*. Capitals draw attention to specific instances of people, places, and things. Some categories that should be capitalized include the following:

- brand names
- companies
- weekdays
- months
- governmental divisions or agencies

- historical eras
- major historical events
- holidays
- institutions
- famous buildings
- ships and other manmade constructions
- natural and manmade landmarks
- territories
- nicknames
- epithets
- organizations
- planets
- nationalities
- tribes
- religions
- names of religious deities
- roads
- special occasions, like the Cannes Film Festival or the Olympic Games

Exceptions

Related to American government, capitalize the noun Congress but not the related adjective congressional. Capitalize the noun U.S. Constitution, but not the related adjective constitutional. Many experts advise leaving the adjectives federal and state in lowercase, as in *federal regulations* or *state water board*, and only capitalizing these when they are parts of official titles or names, like *Federal Communications Commission* or *State Water Resources Control Board*. While the names of the other planets in the solar system are capitalized as names, Earth is more often capitalized only when being described specifically as a planet, like *Earth's orbit*, but lowercase otherwise since it is used not only as a proper noun but also to mean *land, ground, soil*, etc.

Names of animal species or breeds are not capitalized unless they include a proper noun. Then, only the proper noun is capitalized. *Antelope, black bear*, and *yellow-bellied sapsucker* are not capitalized. However, *Bengal tiger, German shepherd, Australian shepherd, French poodle*, and *Russian blue cat* are capitalized.

Other than planets, celestial bodies like the *sun, moon*, and *stars* are not capitalized. Medical conditions like *tuberculosis* or *diabetes* are lowercase; again, exceptions are proper nouns, like *Epstein-Barr syndrome, Alzheimer's disease*, and *Down syndrome*. Seasons and related terms like *winter solstice* or *autumnal equinox* are lowercase. Plants, including fruits and vegetables, like *poinsettia, celery*, or *avocado*, are not capitalized unless they include proper names, like *Douglas fir, Jerusalem artichoke, Damson plums*, or *Golden Delicious apples*.

Titles and Names

When official titles precede names, they should be capitalized, except when there is a comma between the title and name. But if a title follows or replaces a name, it should not be capitalized. For example, "the president" without a name is not capitalized, as in "The president addressed Congress." But with a name it is capitalized, like "President Obama addressed Congress." Or, "Chair of the Board Janet Yellen was appointed by President Obama." One exception is that some publishers and writers nevertheless capitalize President, Queen, Pope, etc., when these are not accompanied by names to show respect for these high offices. However, many writers in America object to this practice for violating democratic principles of equality. Occupations before full names are not capitalized, like owner Mark Cuban, director Martin Scorsese, or coach Roger McDowell.

Some universal rules for capitalization in composition titles include capitalizing the following:

- The first and last words of the title
- Forms of the verb *to be* and all other verbs
- Pronouns
- The word *not*

Universal rules for NOT capitalizing include the articles *the, a,* or *an;* the conjunctions *and, or,* or *nor,* and the preposition *to,* or *to* as part of the infinitive form of a verb. The exception to all of these is UNLESS any of them is the first or last word in the title, in which case they are capitalized. Other words are subject to differences of opinion and differences among various stylebooks or methods. These include *as, but, if,* and *or,* which some capitalize and others do not. Some authorities say no preposition should ever be capitalized; some say prepositions five or more letters long should be capitalized. The *Associated Press Stylebook* advises capitalizing prepositions longer than three letters (like *about, across,* or *with*).

Ellipses
Ellipses (. . .) signal omitted text when quoting. Some writers also use them to show a thought trailing off, but this should not be overused outside of dialogue. An example of an ellipsis would be if someone is quoting a phrase out of a professional source but wants to omit part of the phrase that isn't needed: "Dr. Skim's analysis of pollen inside the body is clearly a myth . . . that speaks to the environmental guilt of our society."

Commas
Commas separate words or phrases in a series of three or more. The **Oxford comma** is the last comma in a series. Many people omit this last comma, but doing so often causes confusion. Here is an example:

> I love my sisters, the Queen of England and Madonna.

This example without the comma implies that the "Queen of England and Madonna" are the speaker's sisters. However, if the speaker was trying to say that they love their sisters, the Queen of England, as well as Madonna, there should be a comma after "Queen of England" to signify this.

Commas also separate two coordinate adjectives ("big, heavy dog") but not cumulative ones, which should be arranged in a particular order for them to make sense ("beautiful ancient ruins").

A comma ends the first of two independent clauses connected by conjunctions. Here is an example:

> I ate a bowl of tomato soup, and I was hungry very shortly after.

Here are some brief rules for commas:

- Commas follow introductory words like however, furthermore, well, why, and actually, among others.

- Commas go between a city and state: Houston, Texas.

- If using a comma between a surname and Jr. or Sr. or a degree like M.D., also follow the whole name with a comma: "Martin Luther King, Jr., wrote that."

- A comma follows a dependent clause beginning a sentence: "Although she was very small, . . ."

- Nonessential modifying words/phrases/clauses are enclosed by commas: "Wendy, who is Peter's sister, closed the window."

- Commas introduce or interrupt direct quotations: "She said, 'I hate him.' 'Why,' I asked, 'do you hate him?'"

Semicolons

Semicolons are used to connect two independent clauses, but should never be used in the place of a comma. They can replace periods between two closely connected sentences: "Call back tomorrow; it can wait until then." When writing items in a series and one or more of them contains internal commas, separate them with semicolons, like the following:

People came from Springfield, Illinois; Alamo, Tennessee; Moscow, Idaho; and other locations.

Hyphens

Here are some rules concerning hyphens:

- Compound adjectives like *state-of-the-art* or *off-campus* are hyphenated.

- Original compound verbs and nouns are often hyphenated, like "throne-sat," "video-gamed," "no-meater."

- Adjectives ending in *–ly* are often hyphenated, like "family-owned" or "friendly-looking."

- "Five years old" is not hyphenated, but singular ages like "five-year-old" are.

- Hyphens can clarify. For example, in "stolen vehicle report," "stolen-vehicle report" clarifies that "stolen" modifies "vehicle," not "report."

- Compound numbers twenty-one through ninety-nine are spelled with hyphens.

- Prefixes before proper nouns/adjectives are hyphenated, like "mid-September" and "trans-Pacific."

Parentheses

Parentheses enclose information such as an aside or more clarifying information: "She ultimately replied (after deliberating for an hour) that she was undecided." They are also used to insert short, in-text definitions or acronyms: "His FBS (fasting blood sugar) was higher than normal." When parenthetical information ends the sentence, the period follows the parentheses: "We received new funds ($25,000)." Only put periods within parentheses if the whole sentence is inside them: "Look at this. (You'll be astonished.)" However, this can also be acceptable as a clause: "Look at this (you'll be astonished)." Although parentheses appear to be part of the sentence subject, they are not, and do not change subject-verb agreement: "Will (and his dog) was there."

Quotation Marks

Quotation marks are typically used when someone is quoting a direct word or phrase someone else writes or says. Additionally, quotation marks should be used for the titles of poems, short stories, songs, articles, chapters, and other shorter works. When quotations include punctuation, periods and commas should *always* be placed inside of the quotation marks.

When a quotation contains another quotation inside of it, the outer quotation should be enclosed in double quotation marks and the inner quotation should be enclosed in single quotation marks. For example: "Timmy was begging, 'Don't go! Don't leave!'" When using both double and single quotation marks, writers will find that many word-processing programs may automatically insert enough space between the single and double quotation marks to be visible for clearer reading. But if this is not the case, the writer should write/type them with enough space between to keep them from looking like three single quotation marks. Additionally, non-standard usages, terms used in an unusual fashion, and technical terms are often clarified by quotation marks. Here are some examples:

My "friend," Dr. Sims, has been micromanaging me again.

This way of extracting oil has been dubbed "fracking."

Apostrophes

One use of the apostrophe is followed by an *s* to indicate possession, like *Mrs. White's home* or *our neighbor's dog*. When using the *'s* after names or nouns that also end in the letter *s*, no single rule applies: some experts advise adding both the apostrophe and the *s*, like "the Jones's house," while others prefer using only the apostrophe and omitting the additional *s*, like "the Jones' house." The wisest expert advice is to pick one formula or the other and then apply it consistently. Newspapers and magazines often use *'s* after common nouns ending with *s*, but add only the apostrophe after proper nouns or names ending with *s*. One common error is to place the apostrophe before a name's final *s* instead of after it: "Ms. Hasting's book" is incorrect if the name is Ms. Hastings.

Plural nouns should not include apostrophes (e.g. "apostrophe's"). Exceptions are to clarify atypical plurals, like verbs used as nouns: "These are the do's and don'ts." Irregular plurals that do not end in *s* always take apostrophe-*s*, not *s*-apostrophe—a common error, as in "childrens' toys," which should be "children's toys." Compound nouns like mother-in-law, when they are singular and possessive, are followed by apostrophe-*s*, like "your mother-in-law's coat." When a compound noun is plural and possessive, the plural is formed before the apostrophe-*s*, like "your sisters-in-laws' coats." When two people named possess the same thing, use apostrophe-*s* after the second name only, like "Dennis and Pam's house."

Usage

Possessives

Possessive forms indicate possession, i.e. that something belongs to or is owned by someone or something. As such, the most common parts of speech to be used in possessive form are adjectives, nouns, and pronouns. The rule for correctly spelling/punctuating possessive nouns and proper nouns is with -*'s*, like "the woman's briefcase" or "Frank's hat." With possessive adjectives, however, apostrophes are not used: these include *my, your, his, her, its, our,* and *their*, like "my book," "your friend," "his car," "her house," "its contents," "our family," or "their property." Possessive pronouns include *mine, yours, his, hers, its, ours,* and *theirs*. These also have no apostrophes. The difference is that possessive adjectives take direct objects, whereas possessive pronouns replace them. For example, instead of using two possessive adjectives in a row, as in "I forgot my book, so Blanca let me use her book," which reads monotonously, replacing the second one with a possessive pronoun reads better: "I forgot my book, so Blanca let me use hers."

Pronouns

There are three pronoun cases: subjective case, objective case, and possessive case. Pronouns as subjects are pronouns that replace the subject of the sentence, such as *I, you, he, she, it, we, they* and *who*.

Pronouns as objects replace the object of the sentence, such as *me, you, him, her, it, us, them,* and *whom.* Pronouns that show possession are *mine, yours, hers, its, ours, theirs,* and *whose.* The following are examples of different pronoun cases:

- Subject pronoun: *She* ate the cake for her birthday. *I* saw the movie.
- Object pronoun: You gave *me* the card last weekend. She gave the picture to *him.*
- Possessive pronoun: That bracelet you found yesterday is *mine. His* name was Casey.

Adjectives

Adjectives are descriptive words that modify nouns or pronouns. They may occur before or after the nouns or pronouns they modify in sentences. For example, in "This is a big house," *big* is an adjective modifying or describing the noun *house.* In "This house is big," the adjective is at the end of the sentence rather than preceding the noun it modifies.

A rule of punctuation that applies to adjectives is to separate a series of adjectives with commas. For example, "Their home was a large, rambling, old, white, two-story house." A comma should never separate the last adjective from the noun, though.

Adverbs

Whereas adjectives modify and describe nouns or pronouns, adverbs modify and describe adjectives, verbs, or other adverbs. Adverbs can be thought of as answers to questions in that they describe when, where, how, how often, how much, or to what extent.

Many (but not all) adjectives can be converted to adverbs by adding *–ly.* For example, in "She is a quick learner," *quick* is an adjective modifying *learner.* In "She learns quickly," *quickly* is an adverb modifying *learns.* One exception is *fast. Fast* is an adjective in "She is a fast learner." However, *–ly* is never added to the word *fast;* it retains the same form as an adverb in "She learns fast."

Verbs

A verb is a word or phrase that expresses action, feeling, or state of being. Verbs explain what their subject is *doing.* Three different types of verbs used in a sentence are action verbs, linking verbs, and helping verbs.

Action verbs show a physical or mental action. Some examples of action verbs are *play, type, jump, write, examine, study, invent, develop,* and *taste.* The following example uses an action verb:

Kat *imagines* that she is a mermaid in the ocean.

The verb *imagines* explains what Kat is doing: she is imagining being a mermaid.

Linking verbs connect the subject to the predicate without expressing an action. The following sentence shows an example of a linking verb:

The mango *tastes* sweet.

The verb *tastes* is a linking verb. The mango doesn't *do* the tasting, but the word *taste* links the mango to its predicate, sweet. Most linking verbs can also be used as action verbs, such as *smell, taste, look, seem, grow,* and *sound.* Saying something *is* something else is also an example of a linking verb. For example, if we were to say, "Peaches is a dog," the verb *is* would be a linking verb in this sentence, since it links the subject to its predicate.

Helping verbs are verbs that help the main verb in a sentence. Examples of helping verbs are *be, am, is, was, have, has, do, did, can, could, may, might, should,* and *must,* among others. The following are examples of helping verbs:

Jessica *is* planning a trip to Hawaii.

Brenda *does* not like camping.

Xavier *should* go to the dance tonight.

Notice that after each of these helping verbs is the main verb of the sentence: *planning, like,* and *go.* Helping verbs usually show an aspect of time.

Transitional Words and Phrases

In connected writing, some sentences naturally lead to others, whereas in other cases, a new sentence expresses a new idea. We use transitional phrases to connect sentences and the ideas they convey. This makes the writing coherent. Transitional language also guides the reader from one thought to the next. For example, when pointing out an objection to the previous idea, starting a sentence with "However," "But," or "On the other hand" is transitional. When adding another idea or detail, writers use "Also," "In addition," "Furthermore," "Further," "Moreover," "Not only," etc. Readers have difficulty perceiving connections between ideas without such transitional wording.

Subject-Verb Agreement

Lack of subject-verb agreement is a very common grammatical error. One of the most common instances is when people use a series of nouns as a compound subject with a singular instead of a plural verb. Here is an example:

Identifying the best books, locating the sellers with the lowest prices, and paying for them *is* difficult

instead of saying "*are* difficult." Additionally, when a sentence subject is compound, the verb is plural:

He and his cousins *were* at the reunion.

However, if the conjunction connecting two or more singular nouns or pronouns is "or" or "nor," the verb must be singular to agree:

That pen or another one like it is in the desk drawer.

If a compound subject includes both a singular noun and a plural one, and they are connected by "or" or "nor," the verb must agree with the subject closest to the verb: "Sally or her sisters go jogging daily"; but "Her sisters or Sally goes jogging daily."

Simply put, singular subjects require singular verbs and plural subjects require plural verbs. A common source of agreement errors is not identifying the sentence subject correctly. For example, people often write sentences incorrectly like, "The group of students *were* complaining about the test." The subject is not the plural "students" but the singular "group." Therefore, the correct sentence should read, "The group of students *was* complaining about the test." The converse also applies, for example, in this incorrect sentence: "The facts in that complicated court case *is* open to question." The subject of the sentence is not the singular "case" but the plural "facts." Hence the sentence would correctly be written: "The facts in that complicated court case *are* open to question." New writers should not be misled by the distance between the subject and verb, especially when another noun with a different number intervenes as in these examples. The verb must agree with the subject, not the noun closest to it.

Pronoun-Antecedent Agreement

Pronouns within a sentence must refer specifically to one noun, known as the **antecedent.** Sometimes, if there are multiple nouns within a sentence, it may be difficult to ascertain which noun belongs to the pronoun. It's important that the pronouns always clearly reference the nouns in the sentence so as not to confuse the reader. Here's an example of an unclear pronoun reference:

> After Catherine cut Libby's hair, David bought her some lunch.

The pronoun in the examples above is *her*. The pronoun could either be referring to *Catherine* or *Libby*. Here are some ways to write the above sentence with a clear pronoun reference:

> After Catherine cut Libby's hair, David bought Libby some lunch.

> David bought Libby some lunch after Catherine cut Libby's hair.

But many times the pronoun will clearly refer to its antecedent, like the following:

> After David cut Catherine's hair, he bought her some lunch.

Formal and Informal Language

Formal language is less personal than informal language. It is more "buttoned-up" and business-like, adhering to proper grammatical rules. It is used in professional or academic contexts, to convey respect or authority. For example, one would use formal language to write an informative or argumentative essay for school or to address a superior. Formal language avoids contractions, slang, colloquialisms, and first-person pronouns. Formal language uses sentences that are usually more complex and often in passive voice. Punctuation can differ as well. For example, **exclamation points (!)** are used to show strong emotion or can be used as an interjection but should be used sparingly in formal writing situations.

Informal language is often used when communicating with family members, friends, peers, and those known more personally. It is more casual, spontaneous, and forgiving in its conformity to grammatical rules and conventions. Informal language is used for personal emails and correspondence between coworkers or other familial relationships. The tone is more relaxed. In informal writing, slang, contractions, clichés, and the first- and second-person are often used.

Practice Questions

Passage I: Humanities

When Nathaniel Hawthorne wrote *The Scarlet Letter* in 1850, he became the first American author to contribute a mature tragic view of life (1) in the literature of the young United States. He developed a tragic vision that is fundamentally dark: man's inevitable moral downfall is determined by the workings of (2) its own heart. Yet Hawthorne's bleak vision also embraces the drama of the human heart, through which he depicts the heart as a redemptive agent, subtly affirming (3) humanities inherent goodness. He likens the human heart to a cavern. To a visitor, the mouth of the cavern is bright with sunlight and flowers, (4) but a few feet in, the light dims and warmth turns to chill; the visitor stumbles first in confusion, then in terror. (5) Further back, a small gleam of light appears and the visitor hurries toward it, to find a scene much like that at the entrance of the cavern, only perfect. This, to Hawthorne, is the depth of (6) human nature; the beauty that lies beyond fear and hopelessness.

The novel relates the suffering of three people—Hester Prynne, Arthur Dimmesdale, and Roger Chillingworth—each possessing a proud heart that leads them to a choice (7) between a natural morality (represented by the passion that derives from human love) and unnatural morality (symbolized by the code of punishment the Puritan community adheres to). Studying the heart in the context of seventeenth-century Puritanism, Hawthorne focuses on the contradictions between the theological and intellectual definitions of sin, according to the Puritans, and the (8) psychological affects of applying them to the reality of their actions and lives. Within the (9) rigid moral setting of the New England Puritan community, he concentrates upon the sin of pride, fostered in the dim, shadowy hollows of the heart and soul. Man is inevitably faced with the destructive force of pride, and subsequently he is involved in the wrong and guilt it produces. He seeks salvation (the Calvinists believed (10) it would be attained after an open confession of sin), and although he recognizes the bond of sin he shares with all men, the penalty he must pay is one of (11) individual, spiritual, psychological and even physical isolation. (12) The lives of Hawthorne's three protagonists can be described as a gradual awakening achieved through transition from deeds to emotions to comprehension of life's greater meaning, a tri-part cadence (13) that can neither be ignored nor avoided. (14) He shows how Hester's stalwart heart adjusts to the tragedy of her illicit love; how Arthur's heart, enfeebled by pride and lust, declines under sustained strain; and how Chillingworth's stagnant heart putrefies through the act of vengeance. Life's tragedy, (15) with the light and shadows that flit through it, does not defeat Hester Prynne's magnificent heart, and this conclusion tempers Hawthorne's generally gloomy novel, lending a glint of light to his dark but not hopeless view of human nature.

1. The best replacement for the underlined portion would be:
    a. NO CHANGE
    b. to
    c. from
    d. with

2. The best replacement for the underlined portion would be:
    a. NO CHANGE
    b. her own heart
    c. his own heart
    d. people's own heart

3. The best replacement for the underlined portion would be:
    a. NO CHANGE
    b. humanities'
    c. humanitys
    d. humanity's

4. The best replacement for the underlined portion would be:
    a. NO CHANGE
    b. but a few feet in the light, dims and warmth turns to chill, the visitor stumbles, first in confusion, then in terror
    c. but a few feet in the light dims and warmth turns to chill, the visitor stumbles first in confusion then in terror
    d. but a few feet in the light dims and warmth turns to chill: the visitor stumbles first in confusion, then in terror

5. The best replacement for the underlined portion would be:
    a. NO CHANGE
    b. In the back
    c. To the back
    d. Farther in

6. The best replacement for the underlined portion would be:
    a. NO CHANGE
    b. human nature. The beauty that lies beyond fear and hopelessness.
    c. human nature: the beauty that lies beyond fear and hopelessness.
    d. human nature (the beauty that lies beyond fear and hopelessness).

7. The best replacement for the underlined portion would be:
    a. NO CHANGE
    b. between a natural morality (represented by the passion that derives from human love) and an unnatural morality (symbolized by the code of punishment the Puritan community adheres to).
    c. between a natural morality represented by the passion that derives from human love and unnatural morality symbolized by the code of punishment the Puritan community adheres to.
    d. between a natural morality—represented by the passion that derives from human love—and unnatural morality, symbolized by the code of punishment the Puritan community adheres to.

8. The best replacement for the underlined portion would be:
   a. NO CHANGE
   b. psychological damage
   c. psychological changes
   d. psychological effects

9. Which choice most accurately conveys the meaning of the underlined word?
   a. Frightening
   b. Strict
   c. Harsh
   d. Dangerous

10. The best replacement for the underlined portion would be:
    a. NO CHANGE
    b. it would be awarded after an open confession of sin
    c. it would be purchased after an open confession of sin
    d. it would be appropriated after an open confession of sin

11. The best replacement for the underlined portion would be:
    a. NO CHANGE
    b. individual spiritual psychological and even physical isolation
    c. individual, spiritual, psychological, and even physical isolation
    d. individual spiritual, psychological, and even physical isolation

12. Which choice best simplifies the underlined statement?
    a. Through doing and feeling, Hawthorne's three protagonists comprehend the meaning of life
    b. When Hawthorne's three protagonists examine their deeds and the resulting emotions, each gradually comes to comprehend life's greater meaning
    c. Hawthorne's three protagonists gradually awaken during a transition from deeds to emotions to a comprehension of life's greater meaning
    d. Hawthorne's three protagonists each experience a gradual awakening as the emotions generated by their actions lead to an understanding of life's greater meaning

13. The best replacement for the underlined portion would be:
    a. NO CHANGE
    b. can either be ignored or avoided
    c. they should not ignore or avoid
    d. is inescapable

14. The paragraph needs a smooth transition from underlined sentence 13 to underlined underlined sentence 14. Which choice would BEST accomplish this this?
    a. For one of the characters, the result is negative, and for two it is positive.
    b. The process of awakening is difficult for all three characters.
    c. Hawthorne intimately analyzes his characters, exposing the spontaneous movements of their hearts.
    d. This is where Hawthorne's tragic vision becomes clear.

15. Where would the underlined portion fit best in the sentence?
    a. NO CHANGE
    b. After *Hester Prynne's magnificent heart*
    c. After *and this conclusion*
    d. After *generally gloomy novel*

Passage II: History

The vast (16) industrialization of Europe which took place between 1760 and 1840 was perhaps the most significant watershed period in the history of Europe. It was a time of astronomical growth and progress for Europeans as a whole, yet it was simultaneously a wretched and dehumanizing period for the majority of European individuals. Many problems—economic, social, and political—were either created or magnified by the Industrial Revolution: (17) problems which threatened European society not with annihilation but with massive change in the form of a social revolution that would shift the distribution of political might and change class structure.

The Industrial Revolution touched and altered almost every aspect of the economic and political life of Europe prior to 1760, which, in turn, (18) changed the existing social order. (19) With roots digging deeply into the past as the thirteenth century, when capitalism and commerce began to develop, industrialization slowly became (20) inevitable. It was aided by a gradual expansion of the market, a demand for more goods by an increasing number of consumers, and the step-by-step freedom of private enterprise from government control. (21) Industrialization began in earnest in the textile industry of Great Britain, and, due to progress in the field of technological innovations, caused a huge upswing in the amount of money necessary to establish a factory. (22) Industry became the new source of wealth, which had formerly been land, but the power remained concentrated within a small group of rich men—the capitalists. The existence of this wealthy class directly contrasted with that of the impoverished working class (people who had once worked the land, until they were forced by industrialization to undertake factory labor in order to survive). The juxtaposition of a small (23) number of people, who held great wealth, with a population for whom extreme and widespread poverty was inescapable was one of the greatest problems the Industrial Revolution created.

The plight of the industrial family was a shameful aspect of the modernization of Europe. Forced to migrate to areas rich in coal, (24) where filthy, violence-ridden cities have sprung up, the people of agrarian communities encountered suffering, despondency, and poverty. (25) Wages were perpetually too low, and employment was always uncertain. Housing was damp, dirty, cramped, and poorly ventilated, leading to sickness and the spread of disease. Death by starvation was not uncommon, as food was scarce and often unfit to eat.

Child labor was another ugly product of the Industrial Revolution. To help supply their families with the bare minimum of food, clothing, and shelter, children as young as five years old were (26) obliged to (27) work thirteen hours a day in factories six days a week. (28) Half starved; unable to find respite from extreme heat and cold; beaten, kicked, and bruised by overseers, these children were blighted by excessive misery.

(29) The problems created or aggravated by the Industrial Revolution—unfair distribution of wealth, squalid living conditions for much of the working class, child labor, and lack of education—forced a moral adaptation on the part of Europeans. And from the new mindset, the middle class emerged.

16. The best replacement for the underlined portion would be:
    a. NO CHANGE
    b. industrialization of Europe, which took place between 1760 and 1840 was perhaps the most significant watershed
    c. industrialization of Europe, which took place between 1760 and 1840, was perhaps the most significant watershed
    d. industrialization of Europe which took place, between 1760 and 1840, was perhaps the most significant watershed

17. The best replacement for the underlined portion would be:
    a. NO CHANGE
    b. problems threatened European society
    c. problems threatening to European society
    d. problems that threatened European society

18. In the context of the essay as a whole, which of the following words is the most accurate substitution for the underlined portion?
    a. transfigured
    b. disrupted
    c. destroyed
    d. transformed

19. The best replacement for the underlined portion would be:
    a. NO CHANGE
    b. With roots, digging deeply into the past as the thirteenth century
    c. With roots digging as deeply into the past as the thirteenth century
    d. With roots that were digging deeply into the past as the thirteenth century

20. The best replacement for the underlined word would be:
    a. NO CHANGE
    b. indelible
    c. inadvertent
    d. inadvisable

21. A transition sentence is needed to improve the flow from the sentence with number 21 to the following sentence. Which of the following sentences best accomplishes the transition?
    a. The demand for large amounts of capital was met by a small group of wealthy investors.
    b. Therefore, the need of large capitalists was predominant.
    c. Capitalists, with their deep pockets, came to the rescue.
    d. The economic theory of supply and demand was born.

22. The underlined sentence is awkward. Which choice below is the clearest rewrite?
    a. Land, formerly the source of wealth, was overshadowed by industry, dominated by rich capitalists.
    b. Capitalists invested in industry, which replaced land as the primary source of wealth. As a result, power was concentrated in the hands of the capitalists.
    c. By investing their capital in industry, the capitalists changed the source of wealth from land to industry and retained most of the power.
    d. Formerly the major source of wealth, land was replaced by industry, and the capitalists held all the power.

23. The best replacement for the underlined portion would be:
    a. NO CHANGE
    b. amount
    c. quota
    d. mass

24. The best replacement for the underlined portion would be:
    a. NO CHANGE
    b. where filthy, violence-ridden, cities have sprung up
    c. where filthy, violence-ridden cities had sprung up
    d. where filthy, violence ridden cities have sprung up

25. The best replacement for the underlined portion would be:
    a. NO CHANGE
    b. Wages were perpetually too low; and employment was always uncertain.
    c. Wages were perpetually too low, and, employment was always uncertain.
    d. Wages were perpetually too low and employment was always uncertain.

26. The best replacement for the underlined portion would be:
    a. NO CHANGE
    b. impounded
    c. apprenticed
    d. conscripted

27. The best replacement for the underlined portion would be:
    a. NO CHANGE
    b. work; thirteen hours a day in factories, six days a week
    c. work, in factories, for thirteen hours a day, with only Sundays off
    d. work in factories for thirteen hours a day, six days a week

28. The punctuation in the underlined portion is technically correct but awkward. Which of the following is the clearest rewrite with correct punctuation?
    a. Half starved, unable to find respite from extreme heat and cold, beaten, kicked, and bruised by overseers, these children were blighted
    b. Half starved, unable to find respite from extreme heat and cold, and, beaten, kicked, and bruised by overseers, these children were blighted
    c. These half-starved children, unable to find respite from extreme heat and cold, were beaten, kicked, and bruised by overseers and blighted
    d. Half starved, these children who were unable to find respite from extreme heat and cold were beaten, kicked, and bruised by overseers, and they were blighted

29. Is there a weakness in this conclusion?
    a. No, it is fine as is.
    b. It does not mention capitalists.
    c. It contains erroneous information.
    d. It contains extraneous information.

30. Is there an outstanding flaw in the essay as a whole?
    a. No, it is fine as is.
    b. The prose is too dramatic.
    c. It does not cover political change.
    d. It is unfair to the capitalists.

Passage III: Natural Sciences

The escalating rate of obesity in the United States is a major health issue facing all healthcare providers and (31) faculties. Today, at Carter Cullen Medical Center, a community hospital in (32) Wayland, Massachusetts, a simple perioperative step—highlighting body mass index (BMI, determined by dividing weight by height) on surgery schedules and preoperative checklists—is saving lives. In 2014, events involving two patients with high (33) BMI's triggered eighteen months of focused teamwork and communication that resulted in the establishment of a groundbreaking early notification protocol. As a consequence, the healthcare professionals at Carter Cullen Medical Center are better equipped to care for patients who are morbidly obese.

Obesity carries many risks, one of which is sleep apnea. (34) Sleep apnea, in turn, causes problems with intubation during surgery and disrupts the administration of anesthetics or oxygen. Additional adipose tissue in the neck area prevents patients from hyperextending their necks properly. During intubation, the anesthesiologist cannot visualize the vocal cords (35) if the patients neck is not hyperextended, and a fiber-optic intubation device must be inserted. But such a device was not regularly available in Carter Cullen's operating rooms; (36) the equipment was stocked only if difficulty was expected. In the first event, an obese patient needed to be intubated, there was unanticipated difficulty with the intubation, and the patient died.

Another life-threatening risk of obesity is rhabdomyolysis. In the second event, an obese patient recovering from prolonged total joint replacement surgery was resting in one position for an (37) unreasonable amount of time. The patient's weight caused muscle fibers (38) to break down, and release a pigment—harmful to the kidneys—into (39) their circulation. Renal failure and death followed.

These two deaths spurred Carter Cullen Medical Center to develop a program to (40) insure that obese patients are safe in the operating room and in post-op recovery. An extensive task force was formed to explore why the deaths occurred and determine preventive precautions.

The result was a BMI awareness tracking procedure that begins with surgery schedulers in doctors' offices, who now include (41) weight and age information for all patients when scheduling a procedure. If a patient's BMI is thirty-five or higher, it is prominently highlighted on the surgery schedule, so that staff know to prepare the operating room (42) accordingly. (43) All operating room beds are now equipped with pressure-relief mattresses, to distribute a patient's body mass over a broader surface area. The nursing quality council received specialized training in caring for high-BMI patients, and new hires take a mandatory electronic learning program that emphasizes recording BMI on confidential surgery schedules. In addition, the task force established a BMI sensitivity and understanding training program.

These measures have been a success. Since 2016, when the quality care project was fully implemented, of one hundred charts tracked, ninety-one percent included the BMI on the surgery schedule, and ninety-six percent included it on the preoperative checklist.

31. Which, if any, of the following choices is the most accurate choice to replace the underlined word?
    a. NO CHANGE
    b. facilities
    c. cardiologists
    d. the American Obesity Association

32. Which, if any, of the following choices is the most accurate choice to replace the underlined words?
    a. NO CHANGE
    b. Wayland Massachusetts
    c. Wayland, Massachusets
    d. a small Massachusetts town

33. Which, if any, of the following choices is the most accurate choice to replace the underlined word?
    a. NO CHANGE
    b. BMIs
    c. body mass indexes
    d. body mass indices

34. Which, if any, of the following choices is the most accurate choice to replace the underlined words?
    a. NO CHANGE
    b. Sleep apnea, in turn, caused problems with intubation during surgery and disrupted the administration of anesthetics
    c. Sleep apnea, in turn, can cause problems with intubation during surgery that disrupt the administration of anesthetics
    d. Sleep apnea, in turn, causes problems with intubation and the administration of anesthetics

35. Which, if any, of the following choices is the most accurate choice to replace the underlined words?
    a. NO CHANGE
    b. if the patient neck
    c. if the patient's neck
    d. if the patients' neck

36. Which, if any, of the following choices is the most accurate choice to replace the underlined words?
    a. NO CHANGE
    b. the equipment was stocked if difficulty only was expected
    c. the equipment only was stocked if difficulty was expected
    d. only the equipment was stocked if difficulty was expected

37. Within the context of this paragraph, which word makes the most sense?
    a. NO CHANGE
    b. outrageous
    c. unacceptable
    d. extended

38. Which choice clarifies the underlined passage and is grammatically correct?
    a. to break down, and release a pigment harmful to the kidneys
    b. to break down, and release a pigment—harmful to the kidneys
    c. to break down and release a pigment harmful to the kidneys
    d. to break down—and release a pigment—harmful to the kidneys—

39. Which of the following words is a grammatically correct replacement for the underlined portion?
   a. NO CHANGE
   b. they're
   c. there
   d. its

40. Which, if any, of the following choices is the most accurate choice to replace the underlined word?
   a. NO CHANGE
   b. ensure
   c. assure
   d. invalidate

41. Which, if any, of the following choices is the most accurate choice to replace the underlined word?
   a. NO CHANGE
   b. weight and gender
   c. weight and height
   d. weight and race

42. What preparation needs to be made for the operating rooms to be prepared "accordingly"?
   a. All the instruments must be thoroughly sterilized.
   b. A fiber-optic intubation device must be on hand.
   c. The operating table must be equipped with a scale.
   d. A back-up anesthesiologist must be in attendance.

43. How do the mattresses mentioned in the underlined sentence help patients?
   a. They make the patients more comfortable than regular mattresses.
   b. They are soft, which helps muscles relax.
   c. They distribute body weight more evenly than regular mattresses, so pressure is minimized.
   d. They massage the body, which makes sleeping easier.

44. Which sentence would make the most logical conclusion to the essay?
   a. No intubation emergencies or positioning incidents have occurred in patients with a high BMI since January 1, 2016.
   b. Doctors, nurses, and administrative staff credit the task force for this huge success.
   c. A cost-benefit task force has been convened to assess the financial impact of the new program.
   d. Carter Cullen Medical Center hopes to share the new program with other hospitals across the country.

45. The writer has been asked to provide a subtitle for this essay. Which choice best summarizes the content?
   a. New Early Notification Protocol at Carter Cullen Medical Center
   b. Hospital Task Force Saves Lives
   c. Local Hospital Has Success with Obese Patients
   d. How One Hospital Prevents Perioperative Complications in Obese Patients

Passage IV: Natural Sciences

(Paragraph 1) Cerebrovascular accident, commonly known as stroke, is the fifth leading cause of death in the United States, and it (46) <u>affects</u> approximately 795,000 Americans each year. (47) <u>The risk of stroke varies by demographics and lifestyle factors such as age (fifty years or older), gender (female), race (African American, American Indian, and Alaskan native), geography (southeastern states), disease, and lifestyle components (including, but not limited to: exercise, eating habits, use of alcohol and/or illicit drugs, cigarette smoking, stress, quality of life, and perceived happiness).</u>

(Paragraph 2) What exactly is a stroke? It is a sudden interruption in cerebral (brain) function that (49) <u>lasts more than twenty-four hours. Stroke can result in death, and is caused by an acute disruption of blood flow coupled with insufficient oxygen reaching the brain.</u> A stroke can be large and catastrophic, ending in death; mild, with the person who had the stroke eventually recovering most normal functions; or anywhere between these two (50) <u>extremes.</u>

(Paragraph 3) The two main types of stroke are ischemic and hemorrhagic. Ischemic stroke occurs when a blood clot blocks an artery in the central nervous system (CNS). (51) <u>The blockage disrupts the flow of oxygen-rich blood to CNS tissue and to the brain to the extent that the supply of oxygen and glucose is insufficient to support ongoing metabolism.</u> Thrombotic stroke and embolic stroke are two subtypes of ischemic stroke. In thrombotic stroke, a blood clot or atherosclerotic plaque develops (52) <u>locally</u> to create blockage. (53) <u>In embolic stroke, a blood clot or atherosclerotic plaque develops elsewhere, such as in the heart, breaks apart, and travels in the bloodstream without blocking blood flow until it reaches a central nervous system artery, then it travels through the artery up to the brain, lodging there.</u>

(Paragraph 4) (54) <u>Headache, nausea, vomiting, numbness, and loss of consciousness are all warning signs of a possible stroke these symptoms require immediate medical attention to prevent damage to—or complete loss of—brain cells.</u> The three most common symptoms of stroke are varying degrees of facial drooping, unilateral limb weakness, and speech difficulty. Noncontrast computed tomography (CT) scans can detect mass lesions, such as a tumor or an abscess, and an acute hemorrhage. Strokes can also cause depression and a loss of the ability to control emotions.

(Paragraph 5) Hemorrhagic stroke is caused when a blood vessel ruptures, sending blood into the surrounding brain tissue. Intracerebral hemorrhage is one subtype of hemorrhagic stroke and is characterized by bleeding within brain tissue. Subarachnoid hemorrhage is a second subtype and occurs when bleeding is within the space between the arachnoid and pia mater membranes of the meninges, the connective tissue around the brain. Bleeding within the brain is a serious condition, (56) <u>as the blood leaked can destroy brain tissue</u> and impairment can happen within a matter of minutes.

(Paragraph 6) With both ischemic stroke and hemorrhagic stroke, structural damage in the central nervous system can disrupt connecting pathways, leading to significant loss of neurologic function and disability. (57) <u>The development of neurological disease may be a consequence of stroke-associated disruption of certain neuronal pathways.</u>

46. Which choice, if any, would be an acceptable substitute for the underlined portion?
    a. destroys
    b. effects
    c. kills
    d. afflicts

47. Carlos is a 27-year-old Hispanic male who lives in New York, and he maintains a healthy lifestyle by eating wholesome foods, exercising regularly, and getting an average of eight hours of sleep per night. He enjoys an occasional beer or glass of wine on weekends, but he does not drink to excess and has never used illicit drugs. He does not smoke, he practices yoga and meditation to control stress, and he is generally satisfied with where he is in his life today. According to the underlined passage, how does this affect Carlos?

    a. He will never have a stroke if he continues living as he does now.
    b. He will never have a stroke because he is not a "match" in any of the categories.
    c. He will have a stroke if he stops exercising, gains weight, and starts smoking cigarettes.
    d. He might be one of the Americans who has a stroke.

48. If this sentence were added at the end of Paragraph 2, what would the effect be? "The prevalence of pseudobulbar affect among stroke survivors is estimated to be between eleven and fifty-two percent."

    a. It would be confusing because the information is irrelevant to the essay.
    b. It would be helpful because it adds information that supports the topic of the paragraph.
    c. It would be confusing because pseudobulbar affect has not been introduced previously.
    d. It would be helpful because it cites a relatively positive stroke survival statistic.

49. What is the most logical ordering of the three items in the underlined passage?

    a. NO CHANGE
    b. is caused by an acute disruption of blood flow, lasts more than twenty-four hours, and can result in death
    c. can result in death, lasts more than twenty-four hours, and is caused by an acute disruption of blood flow
    d. can result in death, is caused by an acute disruption of blood flow, lasts more than twenty-four hours

50. In the context of the sentence, which word is the closest synonym to the underlined word?

    a. Spectrums
    b. States
    c. Poles
    d. Positions

51. Where is the best place to insert a comma in the underlined passage?

    a. The punctuation is correct as is.
    b. The blockage disrupts the flow of oxygen-rich blood to CNS tissue and to the brain, to the extent that the supply of oxygen and glucose is insufficient to support ongoing metabolism.
    c. The blockage disrupts the flow of oxygen-rich blood to CNS tissue, and to the brain to the extent that the supply of oxygen and glucose is insufficient to support ongoing metabolism.
    d. The blockage disrupts the flow of oxygen-rich blood, to CNS tissue and to the brain to the extent that the supply of oxygen and glucose is insufficient to support ongoing metabolism.

52. In the context of this paragraph, what does the word "locally" word mean?

    a. In the brain
    b. In the CNS
    c. Anywhere in the arteries where there are clots and plaque
    d. In the fat cells

53. There is a semicolon missing from this run-on sentence. What is the correct position for it?

  a. In embolic stroke, a blood clot or atherosclerotic plaque develops elsewhere, such as in the heart; breaks apart, and travels in the bloodstream without blocking blood flow until it reaches a central nervous system artery, then it travels through the artery up to the brain, lodging there.

  b. In embolic stroke, a blood clot or atherosclerotic plaque develops elsewhere, such as in the heart, breaks apart, and travels in the bloodstream without blocking blood flow until it reaches a central nervous system artery; then it travels through the artery up to the brain, lodging there.

  c. In embolic stroke, a blood clot or atherosclerotic plaque develops elsewhere; such as in the heart, breaks apart, and travels in the bloodstream without blocking blood flow until it reaches a central nervous system artery, then it travels through the artery up to the brain, lodging there.

  d. In embolic stroke, a blood clot or atherosclerotic plaque develops elsewhere, such as in the heart, breaks apart; and travels in the bloodstream without blocking blood flow until it reaches a central nervous system artery, then it travels through the artery up to the brain, lodging there.

54. Which punctuation mark is the correct one to place between "stroke" and "these" as is?

  a. Slash (/)
  b. Dash (—)
  c. Semicolon (;)
  d. Period (.)

55. Which sentence, if any, interrupts the flow of the discussion in Paragraph 4?

  a. None
  b. The sentence beginning "The three most common symptoms. . ."
  c. The sentence beginning "Noncontrast computed tomography. . ."
  d. The sentence beginning "Strokes can also cause depression. . ."

56. Which is correct?

  a. NO CHANGE
  b. as the blood is leaked can destroy brain tissue
  c. as the blood is leaked, can destroy brain tissue
  d. because as blood leaked, it can destroy brain tissue

57. Which outline of types and subtypes of stroke is correct?

  a.
  I. Ischemic
    A. Thrombotic
    B. Atherosclerotic
  II. Hemorrhagic
    A. Intracerebral
    B. Subarachnoid

  b.
  I. Ischemic
    A. Thrombotic
    B. Embolic
  II. Hemorrhagic
    A. Intracerebral
    B. Subarachnoid

  c.
  I. Hemorrhagic
    A. Thrombotic
    B. Embolic
  II. Ischemic
    A. Intracerebral
    B. Subarachnoid

  d.
  I. Ischemic
    A. Thrombotic
    B. Embolic
  II. Hemorrhagic
    A. Arachnoid
    B. Pia mater

58. The paragraphs in the essay are numbered 1-6. Which ordering structure is most logical?
  a. 1, 2, 3, 5, 6, 4
  b. 1, 2, 3, 4, 6, 5
  c. 1, 2, 4, 3, 5, 6
  d. 1, 4, 2, 3, 5, 6

59. Working with the paragraphs as numbered, in which paragraph would the following information most logically be located? "The pathophysiology of hemorrhagic stroke is primarily attributed to the presence of blood in extracellular tissues of the central nervous system. The primary damage to surrounding brain tissue is related to the mass effect of excess blood on neural structures."
  a. Paragraph number 2
  b. Paragraph number 3
  c. Paragraph number 5
  d. Paragraph number 6

60. Which of the following statements is false?
   a. Approximately 795,000 Americans die of stroke each year.
   b. Thrombotic stroke and embolic stroke are subtypes of ischemic stroke, and they are caused by blood clots.
   c. The primary damage to brain tissue is related to the mass effect of excess blood.
   d. Both types of stroke disrupt the normal activity of structures in the CNS.

Passage V: Social Sciences (Linguistics)

The concept of three-ness, (61) embodied by triplets, triads, and trios, is crucial to Indo-European societies as a way of creating order and stability within the social structure. (62) The significance of the notion of three-ness is manifested in the literature of these cultures, which abounds with numerous events, occurring in patterns of three, that radically affect the lives of the main characters and (63) those of there families, friends, and foes. In no body of literature, perhaps, is this more (64) distinctly and consistently seen than in that of the early Icelandic (65) inhabitants. (66) The sagas produced by the Icelanders are surfeited with images of three, minor and major; it is in the latter, and most widely acclaimed sagas (*The Saga of Gisli, Laxdaela Saga,* and *Njal's Saga*), that three-ness is applied most skillfully.

Saga authors exercised authorial control by using three-ness to create a balanced, rhythmic, organically unified fiction grounded in historical fact. (67) There are two levels of significance in how they applied the concept of three. The first level consists of instances in which the author mentions the number three (or multiples of three) that are minor in import. For example, three is widely used to name the number of sons the saga-man has, or the number of days he spends visiting a friend. Small multiples of three, such as fifteen and eighteen, mark the number of warriors fighting on either side in a skirmish, while greater multiples, such as sixty, 120, and three hundred are predominately associated with wealth or livestock. (68) These recurring instances of three-ness serve a dual purpose. (69) Not only do they create a regular, periodic cadence—a steady underlying tempo—that links all of the sections of the saga together, but they also provide, by means of the (70) tripod-like sense of proportion, an equilibrium associated with the number three: a satisfying feeling of balance.

(71) The third level of relevance concerning the authorial application of three-ness deals with the aesthetic structure of the sagas. It consists of the division of each saga into three sections, plot incidents yoked together in clusters of three, the reenactment of a specific deed or action three times, and the combination of several types of the minor instances of three-ness within the body of one episode that is particularly important in the life of the saga-man. (72) For instance in Part III, Chapter III of the saga, the saga-man might slay an archenemy a third of the way through the third night of his residence at his adversary's abode, three villages away from his own.

The application of imagination to the quantitative and qualitative use of the notion of three-ness did not suddenly appear, fully established and perfected, in the earliest of Icelandic literature. Rather, it was gradually introduced, adapted, modified, and improved upon by many authors until it reached an artistically sophisticated stage. The frequency and significance of the usage of the number three increases in richness as the sagas progress chronologically according to their dates of composition.

61. The definition of the underlined word is "to show in concrete form," and it is not entirely appropriate in this context. Which choice is most accurate?
   a. Marked
   b. Earmarked
   c. Included
   d. Represented

62. Which word or phrase should be eliminated to make the sentence less wordy?
   a. The significance of
   b. the notion of
   c. of these cultures
   d. numerous

63. If there is an error in the underlined sentence, what is the error?
   a. NO ERROR
   b. "Those" should be "that."
   c. "There" should be "their."
   d. There should not be a comma before "and foes."

64. Choose the pair of words that is the closest substitute for "distinctly and consistently."
   a. Discerningly and constantly
   b. Clearly and uniformly
   c. Definitely and evenly
   d. Easily and transparently

65. What is the closest synonym to the underlined word?
   a. Dwellers
   b. Authors
   c. Cultures
   d. Persons

66. This sentence originally appeared at the end of the first paragraph, but it was later deleted: "The sagas are so noticeably saturated with trinities, threesomes, and treble happenings that the modern reader almost comes away from reading them with the numeral three indelibly stamped on her consciousness." What is the best reason for its deletion?
   a. It says the modern reader is female, which excludes the ones who are male.
   b. "Trinities, threesomes, and treble happenings" echoes the first sentence's "triplets, triads, and trios" and seems a bit forced, especially "treble happenings."
   c. What happens to the modern reader is irrelevant.
   d. The writer cannot possibly know the impact the repetition of threes has on anyone but his or her own self.

67. Which sentence would be the clearest, if inserted in this paragraph?
   a. There are two levels of significance to how they applied the concept of three.
   b. How they applied the concept of three has two levels.
   c. They applied the concept of three on two significant levels.
   d. There are two levels of significance to the application of three.

68. The underlined sentence begins a new topic. This sentence and the rest of the paragraph would work best as a separate paragraph. Where should the new paragraph be located?
   a. Immediately after the paragraph it is currently in
   b. Immediately after the first paragraph
   c. Immediately before the last paragraph
   d. Immediately after the last paragraph

69. Which is the clearest way of breaking this sentence into two sentences?

a. They create a regular, periodic cadence—a steady underlying tempo—that links all of the sections of the saga together. They also provide, by means of the tripod-like sense of proportion, an equilibrium associated with the number three: a satisfying feeling of balance.

b. Not only do they create a regular, periodic cadence that links all of the sections of the saga together. They also provide, by means of the tripod-like sense of proportion, an equilibrium associated with the number three.

c. Not only do they create a regular, periodic cadence—a steady underlying tempo—that links all of the sections of the saga together. But they also provide, by means of the tripod-like sense of proportion, an equilibrium associated with the number three: a satisfying feeling of balance.

d. A regular, periodic cadence—a steady underlying tempo—that links all of the sections of the saga together is created. Also, an equilibrium associated with the number three, a satisfying feeling of balance, is provided by means of the tripod-like sense of proportion.

70. What other image might the writer use to describe the "tripod-like" sense of proportion?

a. A drawing compass
b. A tricycle
c. A three-legged dog
d. A three-legged stool

71. Where is the error in the underlined passage?

a. The word "aesthetic" doesn't make sense in this context.
b. The word "chiefly" should be inserted after "deals."
c. There are only two levels of relevance.
d. Saga-men, not authors, apply the concept of three-ness.

72. In this sentence, one comma is missing. Where should it be inserted?

a. After "instance"
b. After "residence"
c. After "Chapter III"
d. After "archenemy"

73. Why was three-ness so important to the early Icelanders?

a. They found it easiest to count and group things, such as livestock, by threes.
b. Three was a significant number in their religion.
c. It was a way to impose order.
d. They needed a way to make their literature unique.

74. What is the major flaw in this essay?

a. There's no mathematical back-up to show why three is so important.
b. It doesn't explain how order and stability are created through the use of multiple threes.
c. The concluding paragraph is off-topic.
d. The examples are too complicated to follow easily.

75. Which idea is not supported in the essay?

a. The concept of three lends a sense of unity and proportion to the sagas.
b. Saga authors demonstrate awareness of the need of an aesthetic element by representing three-ness in the structure of their sagas.
c. Saga authors were interested in both telling a good story and the mechanics of writing.
d. The logical symmetry of the numbers two and four is equal to the balance of three.

Answer Explanations

1. B: The underlined portion of this sentence is a preposition. Prepositions connect an object to nouns, to pronouns, or to a phrase representing a noun. Here, the noun is "view," Hawthorne's tragic view, and the object is "the literature of the United States." The preposition is the word "in," which indicates the vision is already within the body of literature. But the beginning of the sentence makes it clear that Hawthorne's vision is a new addition, so the correct preposition is "to," to show possession (the literature now includes Hawthorne's vision). "From" is incorrect because it is the opposite of "to," and "with" is incorrect because it signals that Hawthorne and the literature are co-contributors.

2. C: Collective nouns denote a group as a single unit, and they typically take a singular pronoun ("Mankind had its finest hour.") In this passage, "man" is taken as the universal, collective subject. "His" agrees with "man" and in this context is grammatically correct. Choice *B* ("her") would make the valid point that traditional constructs unfairly exclude women, but that argument is irrelevant to the discussion of Hawthorne's allegory of the human heart. Choice *D*, people's, is grammatically correct, but it is not the best choice, because the sentence is already about one collective subject, "man," and does not need another collective term, "people."

3. D: Choice *D* is a singular possessive noun. The other choices do not follow correct sentence structure. Choice *A* makes no sense; "humanities" is the study of human culture, which does not possess inherent goodness. Choice *B* is the plural possessive of "humanities." Choice *C*, "humanitys," is not a word.

4. A: The highlighted portion is an example of two independent clauses held together by a semicolon. The first clause is "but a few feet in, the light dims and warmth turns to chill," and the second one is "the visitor stumbles first in confusion, then in terror." The introductory phrase "but a few feet in" is an absolute phrase and must end with a comma. Inserting a comma after "dims" is inappropriate because the conjunction "and" connects only two actions. In the second clause, the comma after "confusion" substitutes for the conjunction "and." Choice *B* is incorrect because the comma separates the noun "light" from its verb "dims." Choice *C* is incorrect because the commas are needed to separate the phrases within the clauses. Choice *D* is incorrect because it lacks the appropriate commas and because the clause following the colon does not introduce a summary or an explanation.

5. D: "Farther" indicates additional distance, whereas "further" indicates an additional amount of something abstract, such as time. Choice *B* interrupts the sense of journey the writer is describing, and therefore is not the best answer. Choice *C* is awkward; "toward" rather than "to" is correct.

6. C: Choice *C* is correct because a colon is used to introduce a summary or an explanation. The phrase "the beauty that lies beyond fear and hopelessness" further describes Hawthorne's vision of human nature. Choice *A* is incorrect because the semicolon is not separating two independent clauses. Choice *B* is incorrect because it is a fragment, not a complete sentence. Choice *D* is incorrect because parentheses are used to enclose nonessential elements such as minor digressions.

7. B: Choice *B* is correct because inserting the article "an" creates parallel construction with "a natural morality." Choices *A*, *C*, and *D* are incorrect because they lack parallel construction.

8. D: The word "affect" is a verb that means "to have an influence on," whereas "effect" in this sentence is a noun synonymous with "result." Choices *B* and *C* are incorrect because they alter the writer's intended meaning.

9. B: While Choices *A*, *C*, and *D* might all describe the moral setting of New England's Puritan community, Choice *B* is the closest synonym to "rigid."

10. A: Choice *A* is correct, because the passage indicates salvation is earned by an individual through right actions. It is not given (Choice *B*), bought (Choice *C*), or taken (Choice *D*).

11. D: The placement of commas changes the meaning in this passage. This passage says a person must pay for their sins by themselves. Therefore the word "individual" modifies an isolation that takes place on spiritual, psychological, and physical planes. Choice *B* is incorrect because without commas, there is no way to be certain of the writer's intention. Choice *C* is incorrect because in it "individual" is used as an adjective, along with "spiritual," "psychological," and "physical."

12. B: Choice *A* is to the point, but in its brevity, it loses the idea of progression. Choice *C* is confusing because there is not enough information given about the awakening itself. Is it a physical awakening, or perhaps a spiritual one? Choice *D* includes all the information in the underlined passage, but is equally wordy. Choice *B* is the best choice because it states the action in a clear, straightforward fashion.

13. A: The coordinating conjunction "nor" pairs with "neither," as "or" pairs with "either." Choice *B* is incorrect because it states the opposite of what is in the underlined passage. Choices *C* and *D* are incorrect because they do not reflect the writer's intention.

14. C: Choices *A*, *B*, and *D* are all true statements, but they do not connect the underlined passage 13 to underlined passage 14. The correct answer is Choice *C* because it describes how Hawthorne arrived at the results reported in underlined passage 14.

15. B: Choice *B* is correct. It could be argued that the conclusion (Choice *C*) and the novel (Choice *D*) both contain light and shadows, but the essay is about Hawthorne's allegory of the human heart as a cave holding both light and shadows; it is most fitting that Hester's heart be described that way.

16. C: The correct answer is Choice *C* because the phrase "which took place between 1760 and 1840" is set off with commas. Choice *B* is incorrect because it lacks a comma after "1840," and Choice *D* is incorrect because the comma after "place" interrupts the phrase "took place between 1760 and 1840."

17. D: Choice *B* is incorrect because the word "that" is missing. The phrase should be "problems *that* threatened European society." Choice *C* is incorrect because it is awkward and confusing. Choice *D* is the correct answer because it uses "that," which does not take a preceding comma, as "which" must.

18. D: Choice *D*, "transformed," is the most accurate substitution because the essay describes the dramatic demographic changes that happened to European society during the Industrial Revolution. Choice *A* ("transfigured") suggests a radical change in figure or appearance, and as such is less exact than "transformed." Choice *B* ("disrupted") does not convey the idea of change, and Choice *C* ("destroyed") is inaccurate because the existing social order was not destroyed.

19. C: The correct answer is Choice *C* because it includes the missing word "as" ("digging *as* deeply"). Choices *A*, *B*, and *D* are incorrect because they do not include the missing word.

20. A: Choice *B*, "indelible," is incorrect because it means permanent. Although industrialization did become permanent, the paragraph is outlining the development of industrialization, not proving its permanence. Choice *C* ("inadvertent") erroneously implies that industrialization was accidental or unintentional. Choice *D* ("inadvisable") classifies industrialization as "unwise," and that thought is not in line with this story of how industrialization developed.

21. A: Choice *A* is the best selection, because it provides the link from the expense of building factories to the capitalists, which led to industry becoming the new source of wealth. Choice *B* is awkwardly worded, and although it correctly states the need for capitalists, it does not put them into action. Choice *C* is factually correct, but is not the best choice because the tone is more casual than that of the rest of the essay. Choice *D* is factually incorrect and contradicts information supplied earlier in the paragraph: supply, demand, and markets existed before the Industrial Revolution.

22. B: Choice *B* is the clearest rewrite. Forming two sentences allows for linear sequencing of the activities. Choice *A* loses information by not including that land was dominated by rich capitalists. Choice *C* is less clear because it refers to the capitalists by "their" before introducing them. Choice *D* is incorrect because it halts the flow from the previous sentence.

23. A: Choice *A* is the correct answer. Choice *B*, "amount" refers to items that cannot be counted, whereas "number" refers to something, such as people, that can be counted. Choice *C*, "quota," refers to a controlled quantity of something one must acquire or receive. Choice *D* is incorrect because "mass" can be used to refer to a large quantity of people, not the relatively small number from the passage.

24. C: Choice *C* is the correct answer because it uses past tense ("had"), which is consistent with the rest of the paragraph. Choice *A* is incorrect because "have" is the present tense, while the passage uses past tense. Choice *B* is incorrect because it inserts an additional comma after "violence-ridden." Choice *D* is incorrect because it leaves out the hyphen on the compound adjective "violence-ridden."

25. D: In a simple sentence such as the underlined passage, there is no need for a comma before the coordinating conjunction "and." Choice *B* is incorrect because with the semicolon, the second half of the sentence becomes a fragment. Choice *C* is incorrect because the conjunction is set off by two commas.

26. A: The correct answer is Choice *A*. Choice *B* ("impounded") indicates the children were seized and confined. If the children were apprenticed (Choice *C*), they would have been legally bound to a tradesman or craftsman to work for a specific amount of time in exchange for learning a skill. If they were conscripted (Choice *D*), they would have been forced into military service.

27. D: The children were obliged to work in factories for thirteen hours a day, six days a week. Choice *A* is incorrect because the sentence is missing a preposition before "thirteen hour a day." Choice *B* is incorrect because it inserts a semicolon but isn't separating independent clauses. Choice *C* is incorrect because it assumes that the day off was always a Sunday, which is not stated elsewhere in the passage.

28. D: Choice *A* is incorrect because "beaten, kicked and bruised by overseers" is one item of the list of cruelties the children faced, yet the placement the phrase blurs it together with "heat and cold." Choice *B* is incorrect because the conjunction "and" is set off by two commas. Choice *C* is incorrect because a comma should follow "overseers," to enclose the treatment at the hands of the overseers as one item. Choice *D* is correct because incorporating the "extreme heat and cold" item into the main clause of the sentence allows "beaten, kicked, and bruised by overseers" to stand as one item delineated by commas.

29. D: The summary lists lack of education as a problem that was either created or aggravated by the Industrial Revolution, but there is no mention of education in the essay.

30. C: The introduction lists three problem areas—economic, social, and political—but the essay discusses only economic and social challenges.

31. B: While cardiologists (Choice *C*) and the American Obesity Association (Choice *D*) are actively dealing with the obesity epidemic in the United States, this paragraph focuses on a hospital's response.

The correct answer is Choice *B*, "facilities." Choice *A* is incorrect because the closest definition of "faculties" that might fit in this context is "teachers and instructors in branches of learning in a college or university." If the writer had intended to use "faculties," they would have identified which branch, such as "medical faculties."

32. A: The spelling and punctuation are correct as written (Choice *A*). Choice *B* lacks a comma between the town and the state, as well as a comma after the state, to enclose the clause. In Choice *C*, Massachusetts is misspelled, and if Choice *D* were used, the essay would then contain less factual information than the original.

33. B: "BMI's" is incorrect because it is the possessive form of "BMI." Choice *B* ("BMIs") is correct because it is the plural form of "BMI." Choices *C* and *D* are incorrect because the acronym is spelled out in full; as it has already been introduced to the reader, it should be used here, not the full term.

34: C: Choice *C* is the correct answer because it states that sleep apnea *can* cause problems; the other choices imply that it always causes problems. In addition, Choices *B* and *D* each have an additional error. Choice *B* is written in the incorrect tense—it uses past tense instead of present tense. Choice *D* omits mentioning that the problems are caused during surgery.

35. C: Choice *A* is incorrect because in the underlined phrase "patients" is plural, indicating more than one patient. Choice *B* is incorrect because "patient" is positioned as a descriptive adjective, rather than as a noun. Choice *D* is partly correct because "patients'" is possessive, but it is incorrect because the word is in the plural form. Choice *C* is correct: it is the singular, possessive form of "patient."

36. A: This question involves proper placement of the adverb "only," which should precede the noun/verb combination it modifies, and it should also be as close to the combination as possible. The underlined phrase, as written, is correct. The device was not in the operating room on a regular basis; one was stocked if the staff knew in advance that the patient had a high BMI. The action being modified is the expectation of difficulty, therefore "only" precedes it and is as close as possible to the words "if difficulty was expected." Choice *B* is incorrect because the adverb follows the action statement. Choice *C* is incorrect because, by following directly behind "equipment," the word "only" attaches itself to that noun. Choice *D* is incorrect because the sentence does not make sense with that wording.

37. D: The correct answer is Choice *D*, "extended." Choices *A* ("unreasonable"), *B* ("outrageous"), and *C* ("unacceptable") are subjective and imply value judgments. "Extended," on the other hand, is factual and tells the reader specifically what it is about the amount of time that leads to rhabdomyolysis.

38. C: The comma before the conjunction "and" in Choice *A* makes this phrase grammatically incorrect. The same is true of Choice *B*. Choice *D* is incorrect because the dashes give the writing a jerky quality. Choice *C* is a simple statement of fact, does not contain a comma before the conjunction, and does not even need dashes.

39. A: A personal pronoun is needed in this phrase, and "their" (Choice *A*) is the correct one. As the reader is not told the patient's gender, the personal pronouns "his" and "her" would be incorrect, and "it" (Choice *D*) would not apply to a person. "They're" (Choice *B*) is a contraction of "they are," which does not make sense in this sentence. Choice *C* ("there") is an adverb, meaning "at, in, or to a place," which also makes no sense.

40. B: Choice *B*, "ensure" is the correct word. "Insure" (Choice *A*) relates to matters of legal and financial protection; "assure" (Choice *C*) means "to promise"; and "invalidate" (Choice *D*) contradicts what the writer is trying to say.

41. C: Choice *C*, "weight and height" is correct. The second sentence in the essay explains that BMI is determined by dividing height into weight.

42. B: The second paragraph discusses the special fiber-optic device that is needed in case there is a problem with intubation or delivering anesthesia, so the correct answer is Choice *B*. The other choices are not listed as necessary preparations in the passage.

43. C: The mattresses are called "pressure-relief" mattresses, which indicates their job is to address the problem of pressure, not patient comfort (Choice *A*), muscle relaxation (Choice *B*), or sleep (Choice *D*).

44. A: Choice *A* is the correct answer because while summarizing the paragraph, it also points back to the opening statement. Choice *B* is redundant in that the essay has already stated that the task force established the new protocol. Choice *C* initiates a new topic, which is the opposite of what a conclusion should do. Choice *D* states what should be an obvious reaction.

45. D: Choice *D* is best because it states three key points clearly and succinctly. Choice *A* is not the best choice because readers outside of the medical profession might not know what an early notification protocol is; Choice *B* is a bit misleading because it's the protocol, not the task force, that saves lives; and Choice *C* is vague because it does not specify what the success is or how it is being achieved.

46. D: "Afflicts" is an acceptable substitute for "affects," in the context of this sentence. Choices *A* ("destroys") and *C* ("kills") are both incorrect choices because not everyone who is affected by a stroke is killed by it. Choice *B* ("effects") is not correct because it is a noun.

47. D: The key phrase in the underlined passage is "*risk* of stroke." The demographics show who in the population is prone to stroke, but they don't guarantee that those people will have strokes. Likewise, they don't guarantee that anyone falling outside of the categories listed will not have a stroke. The correct answer is Choice *D*: It is possible that Carlos might have a stroke in his lifetime.

48. C: Inserted at the end of the paragraph, the sentence about pseudobulbar affect is confusing because the condition has not been introduced yet, and so the reader has no context to put the comment in. The paragraph ends without any additional information that would justify the sentence being there. Choice *A* is incorrect because the information might be very relevant to the essay—but only if it is included in the correct place. Choice *B* is incorrect because the additional information does not support the topic sentence, "What exactly is a stroke?" Choice *D* is incorrect not because the statistic is relatively positive, but because stroke survival is not the topic of the sentence.

49. B: The logical order is the order in which the events occur, as presented in Choice *B*. The blood flow disruption begins, it can be disrupted for up to twenty-four hours, and death can be a result. The other choices do not represent a chronological order of events.

50. C: Choice *C*, "poles," is the closest synonym because as with extremes, poles lie at either end of the spectrum. Choice *A* ("spectrums") is incorrect because there is only one spectrum and it is the continuum of all possibilities that lie between the extremes, including the extremes. Choices *B* ("states") and *D* ("positions") do not connote the sense of opposites that "extremes" does.

51. B: Choice *B* is correct because the comma separates the main clause in the sentence from the subordinating clause. Choice *A* is incorrect because it leaves out a necessary comma after "to the brain." Choice *C* would be correct if there was a comma after "brain" as well. Choice *D* is incorrect because the comma interrupts the prepositional phrase "to CNS tissue."

52. B: The correct answer is Choice *B*. The paragraph discusses clots and plaque that form in the CNS, and clots and plaque that form elsewhere. "Locally" indicates a position in the CNS itself, whereas "a blood clot or atherosclerotic plaque develops elsewhere, such as the heart" indicates a position outside of the CNS.

53. B: A semicolon separates main clauses that are not joined by a coordinating conjunction, such as "and." Choice *B* breaks this long sentence into two main clauses, each of which can stand on its own as a complete sentence.

54. C: The correct answer is Choice *C*, the semicolon. Because there is a set of dashes later in the sentence, Choice *B* ("dash") won't work because dashes should be used sparingly and there's already a pair in the sentence. Choice *A* ("slash") is used to separate two options, not two parts of a sentence. A period, Choice *D*, would be appropriate only if the lowercase "t" in "there" appeared as a capital "T" in the sentence.

55. C: The sentence beginning, "Noncontrast computed tomography. . ." disrupts the flow of the paragraph by inserting a comment about imaging technology into a discussion of stroke symptoms.

56. A: The sentence is correct as written. Choice *B* would be correct only if a comma and the word "it" were inserted after "leaked" ("as the blood is leaked, it can destroy brain tissue"). Choice *C* ("as the blood is leaked, can destroy brain tissue") is incorrect because the pronoun "it" is missing before "can." Choice *D* is incorrect because the tense changes mid-sentence.

57. B: The outline of types and subtypes of stroke is:

      I. Ischemic
            A. Thrombotic
            B. Embolic
      II. Hemorrhagic
            A. Intracerebral
            B. Subarachnoid

58. C: The logical order is:

> 1: Introduction of stroke and stroke demographics

> 2: Definition of stroke

> 4: Symptoms

> 3: Two main types of stroke (ischemic and hemorrhagic); ischemic described

> 5: Hemorrhagic described

> 6: Structural damage in ischemic and hemorrhagic stroke

Choice *A* introduces ischemic and hemorrhagic stroke. It should discuss the types in that order, but it covers hemorrhagic first. Choice *B* discusses structural damage of both types of stroke before covering hemorrhagic stroke in detail. Choice *D* discusses demographics and symptoms of stroke before defining what stroke is, and so the definition of the topic is delivered three paragraphs into the essay.

59. C: The passage to be inserted discusses the pathophysiology of hemorrhagic stroke, so it belongs in the paragraph devoted entirely to hemorrhagic stroke, Paragraph 5 (Choice *C*).

60. A: Choice *A* is false; it states that every American who has a stroke (795,000) dies from it.

61. D: "Represented" is most accurate word. "Marked" (Choice *A*) describes what the triplets, triads, and trios do, rather than what they are—they mark, or flag, the concept of three-ness. Choice *B* ("earmarked") means "marked with an identifying symbol" and is an interesting choice, but not one that is as synonymous as "represented." Choice *C* ("included") is not grammatically correct.

62. B: "The notion of" should be deleted, because "The significance of three-ness. . ." is much more direct. "Of these cultures" (Choice *C*) is a necessary reference to the Indo-European societies. "Numerous" (Choice *D*) attests to the quantity of these events, and it is stronger than simply "abounds with events."

63. C: The error lies in Choice *C*. "Their" is the correct spelling of the possessive form of "they."

64. B: "Clearly and uniformly" is the closest substitute. Choice *A*, "discerningly and constantly," bears little similarity to the original pair aside from beginning with the same letters. Choice *C*, "definitely and evenly," is not immediately clear and would require explanation. In Choice *D*, "easily" is not a synonym of "distinctly."

65. A: "Dwellers" is the closest synonym to "inhabitants." Both are plural nouns and mean "people who reside in or inhabit a certain place." "Authors" (Choice *B*) are a subset of people who might be inhabiting a place; "cultures" (Choice *C*) encompass more than just the people themselves living somewhere; and "persons" (Choice *D*) applies to a specific, small number of people.

66. B: Choice *B* is the correct answer. Style and tone are important aspects of essay writing, and writing that distracts the reader from the point being made is to be avoided. Readers know when a writer is trying too hard, as is the case here.

67. D: This sentence has three pieces of information: there are *two levels of significance* to the *concept of three-ness* that is *applied by the authors*. The reader already knows that the authors of the sagas are the people who are applying the concept of three, so that piece of information does not need to be repeated. Choice *D* is the only choice that omits mentioning the authors and has no flaws. Choice *A* erroneously replaces "in" with "to," which does not make the sentence any clearer. Choice *B* indicates that the application, not the significance, has two levels. Choice *C* is close to the correct answer, but it would be simpler if it did not include the authors.

68. C: If the new paragraph appeared before the final paragraph, the order of topics would flow logically:

- Introduction of concept of three
- First level of significance
- Second level of significance
- Dual purpose of three-ness
- How the application of three-ness evolved over time

69. A: Choice *A* is the clearest rewrite. It replaces the "not only . . . but also. . ." structure with two sentences that are relatively simple to comprehend. Choice *B*, in the interest of brevity, leaves out two concepts: tempo and balance. In Choice *C*, the conjunction "but," which means "on the contrary," throws the intended meaning of the passage into question. Choice *D* is written in the passive voice (the subject is the recipient of the action), which is less direct than the active voice.

70. D: A three-legged stool is the most appropriate of the choices. Choice *A* (drawing compass) won't work because drawing compasses have only two legs. Choice *B* (tricycle) connotes movement along with balance, and in addition to being anachronistic, the image might be a bit childish in the context of this essay. Choice *D*, a three-legged dog, is inappropriate because dogs are not naturally born with just three legs, and it is a tasteless analogy as well.

71. C: Choice *C* is correct. The sentence states that there are three levels of relevance, when in reality there are only two in the passage. Choice *A* is incorrect because "aesthetic," in this sense, is referring to the arrangement of events in the sagas. Choice *B*, insertion of "chiefly" after "deals," is an unnecessary specification, since the paragraph already covers just that. Choice *D* is incorrect because it is the authors who apply the concept of three-ness, not the characters of the sagas.

72. A: The introductory phrase "For instance" should be followed by a comma. The other choices insert unnecessary commas that interrupt the flow of the sentence and decrease clarity.

73. C: This information is imparted in the thesis sentence. Choices *A*, *B*, and *D* all include information not mentioned by the passage. The passage does not indicate early Icelanders found it easier to count in threes (Choice *A*), nor does it say that three was significant to their religion (Choice *B*) or that they felt a need to make their literature unique (Choice *D*).

74. B: The essay fails to support, through explanation and examples, one of the tenets of the thesis statement. Choice *A* is a strong choice, but the essay does not attempt to explore any mathematical significance of three-ness. Choice *C* is incorrect because the conclusion remains on the topic of the three-ness in early Icelandic literature. The examples provided by the passage use simple language, so Choice *D* is also incorrect.

75. D: The only numbers that were discussed as special to the saga authors were three and its multiples. The logic and symmetry of two and four are not discussed in the essay.

Mathematics Test

Preparing for Higher Mathematics

Number & Quantity

Real and Complex Number Systems

Whole numbers are the numbers 0, 1, 2, 3, ... Examples of other whole numbers would be 413 and 8,431. Notice that numbers such as 4.13 and $\frac{1}{4}$ are not included in whole numbers. **Counting numbers**, also known as **natural numbers**, consist of all whole numbers except for the zero. In set notation, the natural numbers are the set $\{1, 2, 3, ...\}$. The entire set of whole numbers and negative versions of those same numbers comprise the set of numbers known as **integers.** Therefore, in set notation, the integers are $\{..., -3, -2, -1, 0, 1, 2, 3, ...\}$. Examples of other integers are −4,981 and 90,131. A number line is a great way to visualize the integers. Integers are labeled on the following number line:

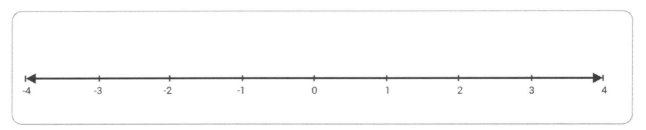

The arrows on the right- and left-hand sides of the number line show that the line continues indefinitely in both directions.

Fractions also exist on the number line and are considered parts of a whole. For example, if an entire pie is cut into two pieces, each piece is half of the pie, or $\frac{1}{2}$. The top number in any fraction, known as the **numerator,** defines how many parts there are. The bottom number, known as the **denominator**, states how many pieces the whole is divided into. Fractions can also be negative or written in their corresponding decimal form.

A **decimal** is a number that uses a decimal point and numbers to the right of the decimal point representing the part of the number that is less than 1. For example, 3.5 is a decimal and is equivalent to the fraction $\frac{7}{2}$ or the mixed number $3\frac{1}{2}$. The decimal is found by dividing 2 into 7. Other examples of fractions are $\frac{2}{7}$, $\frac{-3}{14}$, and $\frac{14}{27}$.

Any number that can be expressed as a fraction is known as a **rational number.** Basically, if a and b are any integers and $b \neq 0$, then $\frac{a}{b}$ is a rational number. Any integer can be written as a fraction where the denominator is 1; therefore, the set of rational numbers consist of all fractions and all integers.

Any number that is not rational is known as an **irrational number.** Consider the number $\pi = 3.141592654$ .... The decimal portion of that number extends indefinitely. In that situation, the number can never be written as a fraction. Another example of an irrational number is $\sqrt{2} = 1.414213662$ .... Again, this number cannot be written as a ratio of two integers.

Together, the set of all rational and irrational numbers makes up the **real numbers.** The number line contains all real numbers. To graph a number other than an integer on a number line, the number will need to be plotted between two integers. For example, 3.5 would be plotted halfway between 3 and 4.

Even numbers are integers that are divisible by 2. For example, 6, 100, 0, and −200 are all even numbers. **Odd numbers** are integers that are not divisible by 2. If an odd number is divided by 2, the result is a fraction. For example, −5, 11, and −121 are odd numbers.

Prime numbers consist of natural numbers greater than 1 that are not divisible by any other natural numbers other than themselves and 1. For example, 3, 5, and 7 are prime numbers. If a natural number is not prime, it is known as a **composite number**. 8 is a composite number because it is divisible by both 2 and 4, which are natural numbers other than itself and 1.

The **absolute value** of any real number is the distance from that number to 0 on the number line. The absolute value of a number can never be negative. For example, the absolute value of both 8 and −8 is 8 because they are both 8 units away from 0 on the number line. This is written as:

$$|8| = |-8| = 8$$

Performing Arithmetic Operations with Rational Numbers
The four basic operations include addition, subtraction, multiplication, and division. The result of addition is a **sum**, the result of subtraction is a **difference**, the result of multiplication is a **product**, and the result of division is a **quotient**. Each type of operation can be used when working with rational numbers; however, the basic operations need to be understood first while using simpler numbers before working with fractions and decimals.

Performing these operations should first be learned using whole numbers. Addition needs to be done column by column. To add two whole numbers, add the ones column first, then the tens columns, then the hundreds, etc. If the sum of any column is greater than 9, a one must be carried over to the next column. For example, the following is the result of 482 + 924:

$$\begin{array}{r} 1 \\ 482 \\ +924 \\ \hline 1406 \end{array}$$

Notice that the sum of the tens column was 10, so a one was carried over to the hundreds column. Subtraction is also performed column by column. Subtraction is performed in the ones column first, then the tens, etc. If the number on top is less than the number below, a one must be borrowed from the column to the left. For example, the following is the result of 5,424 − 756:

$$\begin{array}{r} 4\ 13\ 11\ 14 \\ \cancel{5}\ \cancel{4}\ \cancel{2}\ \cancel{4} \\ -\ 7\ \ 5\ \ 6 \\ \hline 4\ \ 6\ \ 6\ \ 8 \end{array}$$

Notice that a one is borrowed from the tens, hundreds, and thousands place. After subtraction, the answer can be checked through addition. A check of this problem would be to show that 756 + 4,668 = 5,424.

Multiplication of two whole numbers is performed by writing one on top of the other. The number on top is known as the **multiplicand,** and the number below is the **multiplier**. Perform the multiplication by multiplying the multiplicand by each digit of the multiplier. Make sure to place the ones value of each result under the multiplying digit in the multiplier. Each value to the right is then a 0. The product is found by adding each product. For example, the following is the process of multiplying 46 times 37 where 46 is the multiplicand and 37 is the multiplier:

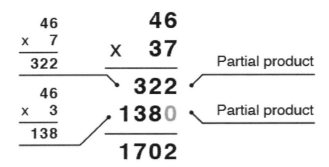

Finally, division can be performed using long division. When dividing a number by another number, the first number is known as the **dividend,** and the second is the **divisor.** For example, with $a \div b = c$, a is the dividend, b is the divisor, and c is the quotient. For long division, place the dividend within the division symbol and the divisor on the outside.

For example, with $8,764 \div 4$, refer to the first problem in the diagram below. First, there are 2 4's in the first digit, 8. This number 2 gets written above the 8.

Then, multiply 4 times 2 to get 8, and that product goes below the 8. Subtract to get 8, and then carry down the 7. Continue the same steps.

$7 \div 4 = 1$ R3, so 1 is written above the 7, and 3 is the remainder.

Multiply 4 times 1 to get 4, and write it below the 7. Subtract to get 3, and carry the 6 down next to the 3. Resulting steps give a 9 and a 1.

The final subtraction results in a 0, which means that 8,764 is divisible by 4. There are no remaining numbers.

The second example shows that:

$$4,536 \div 216 = 21$$

The steps are a little different because 216 cannot be contained in 4 or 5, so the first step is placing a 2 above the 3 because there are 2 216's in 453.

Finally, the third example shows that:

$$546 \div 31 = 17 \, R19$$

The 19 is a remainder. Notice that the final subtraction does not result in a 0, which means that 546 is not divisible by 31. The remainder can also be written as a fraction over the divisor to say that:

$$546 \div 31 = 17\frac{19}{31}$$

```
    2191              21              17 r 19
  4|8764         216|4536         31|546
    8               432             31
    07              216             236
     4              216             217
     36               0              19
     36
      04
       4
       0
```

If a division problem relates to a real-world application, and a remainder does exist, it can have meaning. For example, consider the third example, $546 \div 31 = 17 \, R19$. Let's say that we had \$546 to spend on calculators that cost \$31 each, and we wanted to know how many we could buy. The division problem would answer this question. The result states that 17 calculators could be purchased, with \$19 left over. Notice that the remainder will never be greater than or equal to the divisor.

Once the operations are understood with whole numbers, they can be used with integers. There are many rules surrounding operations with negative numbers. First, consider addition with integers. The sum of two numbers can first be shown using a number line.

For example, to add $-5 + (-6)$, plot the point -5 on the number line. Then, because a negative number is being added, move 6 units to the left. This process results in landing on -11 on the number line, which is the sum of -5 and -6. If adding a positive number, move to the right. Visualizing this process using a number line is useful for understanding; however, it is not efficient. A quicker process is to learn the rules.

When adding two numbers with the same sign, add the absolute values of both numbers, and use the common sign of both numbers as the sign of the sum.

For example, to add $-5 + (-6)$, add their absolute values $5 + 6 = 11$. Then, introduce a negative number because both addends are negative. The result is -11.

To add two integers with unlike signs, subtract the lesser absolute value from the greater absolute value, and apply the sign of the number with the greater absolute value to the result.

For example, the sum $-7 + 4$ can be computed by finding the difference $7 - 4 = 3$ and then applying a negative because the value with the larger absolute value is negative. The result is -3.

Similarly, the sum $-4 + 7$ can be found by computing the same difference but leaving it as a positive result because the addend with the larger absolute value is positive. Also, recall that any number plus 0 equals that number.

This is known as the **Addition Property of 0.**

Subtracting two integers can be computed by changing to addition to avoid confusion. The rule is to add the first number to the opposite of the second number. The opposite of a number is the number on the other side of 0 on the number line, which is the same number of units away from 0. For example, -2 and 2 are opposites. Consider $4 - 8$. Change this to adding the opposite as follows: $4 + (-8)$.

Then, follow the rules of addition of integers to obtain -4. Secondly, consider $-8 - (-2)$. Change this problem to adding the opposite as $-8 + 2$, which equals -6. Notice that subtracting a negative number functions the same as adding a positive number.

Multiplication and division of integers are actually less confusing than addition and subtraction because the rules are simpler to understand. If two factors in a multiplication problem have the same sign, the result is positive. If one factor is positive and one factor is negative, the result, known as the **product,** is negative.

For example:

$$(-9)(-3) = 27$$

and

$$9(-3) = -27$$

Also, any number times 0 always results in 0. If a problem consists of more than a single multiplication, the result is negative if it contains an odd number of negative factors, and the result is positive if it contains an even number of negative factors.

For example:

$$(-1)(-1)(-1)(-1) = 1$$

And

$$(-1)(-1)(-1)(-1)(-1) = 1$$

These two examples of multiplication also bring up another concept. Both are examples of repeated multiplication, which can be written in a more compact notation using exponents. The first example can be written as $(-1)^4 = 1$, and the second example can be written as $(-1)^5 = -1$. Both are exponential expressions, -1 is the base in both instances, and 4 and 5 are the respective exponents.

Note that a negative number raised to an odd power is always negative, and a negative number raised to an even power is always positive.

Also, $(-1)^4$ is not the same as -1^4. In the first expression, the negative is included in the parentheses, but it is not in the second expression. The second expression is found by evaluating 1^4 first to get 1 and then by applying the negative sign to obtain -1.

A similar theory applies within division. First, consider some vocabulary. When dividing 14 by 2, it can be written in the following ways:

$$14 \div 2 = 7 \text{ or } \frac{14}{2} = 7$$

14 is the **dividend,** 2 is the **divisor,** and 7 is the **quotient.**

If two numbers in a division problem have the same sign, the quotient is positive. If two numbers in a division problem have different signs, the quotient is negative.

For example:

$$14 \div (-2) = -7$$

and

$$14 \div (-2) = 7$$

To check division, multiply the quotient by the divisor to obtain the dividend. Also, remember that 0 divided by any number is equal to 0. However, any number divided by 0 is undefined. It just does not make sense to divide a number by 0 parts.

If more than one operation is to be completed in a problem, follow the Order of Operations. The mnemonic device, PEMDAS, for the order of operations states the order in which addition, subtraction, multiplication, and division needs to be done. It also includes when to evaluate operations within grouping symbols and when to incorporate exponents. PEMDAS, which some remember by thinking "please excuse my dear Aunt Sally," refers to parentheses, exponents, multiplication, division, addition, and subtraction. First, within an expression, complete any operation that is within parentheses, or any other grouping symbol like brackets, braces, or absolute value symbols. Note that this does not refer to the case when parentheses are used to represent multiplication like $(2)(5)$ because in such cases, an operation is not within parentheses like it is in $(2 \cdot 5)$. Then, any exponents must be computed. Next, multiplication and division are performed from left to right. Finally, addition and subtraction are performed from left to right.

The following is an example in which the operations within the parentheses need to be performed first, so the order of operations must be applied to the exponent, subtraction, addition, and multiplication within the grouping symbol:

$$9 - 3(3^2 - 3 + 4 \cdot 3)$$

$$9 - 3(3^2 - 3 + 4 \cdot 3) \quad \text{Work within the parentheses first}$$

$$= 9 - 3(9 - 3 + 12)$$

$$= 9 - 3(18)$$

$$= 9 - 54$$

$$= -45$$

Once the rules for integers are understood, move on to learning how to perform operations with fractions and decimals. Recall that a rational number can be written as a fraction and can be converted to a decimal through division. If a rational number is negative, the rules for adding, subtracting, multiplying, and dividing integers must be used. If a rational number is in fraction form, performing addition, subtraction, multiplication, and division is more complicated than when working with integers. First, consider addition. To add two fractions having the same denominator, add the numerators and then reduce the fraction. When an answer is a fraction, it should always be in lowest terms. **Lowest terms** means that every common factor, other than 1, between the numerator and denominator is divided out. For example:

$$\frac{2}{8} + \frac{4}{8} = \frac{6}{8} = \frac{6 \div 2}{8 \div 2} = \frac{3}{4}$$

Both the numerator and denominator of $\frac{6}{8}$ have a common factor of 2, so 2 is divided out of each number to put the fraction in lowest terms. If denominators are different in an addition problem, the fractions must be converted to have common denominators. The **least common denominator (LCD)** of all the given denominators must be found, and this value is equal to the **least common multiple (LCM)** of the denominators. This non-zero value is the smallest number that is a multiple of both denominators. Then, rewrite each original fraction as an equivalent fraction using the new denominator. Once in this form, apply the process of adding with like denominators.

For example, consider $\frac{1}{3} + \frac{4}{9}$.

The LCD is 9 because it is the smallest multiple of both 3 and 9. The fraction $\frac{1}{3}$ must be rewritten with 9 as its denominator. Therefore, multiply both the numerator and denominator by 3. Multiplying by $\frac{3}{3}$ is the same as multiplying by 1, which does not change the value of the fraction. Therefore, an equivalent fraction is:

$$\frac{3}{9}$$

and

$$\frac{1}{3} + \frac{4}{9} = \frac{3}{9} + \frac{4}{9} = \frac{7}{9}$$

which is in lowest terms. Subtraction is performed in a similar manner; once the denominators are equal, the numerators are then subtracted. The following is an example of addition of a positive and a negative fraction:

$$-\frac{5}{12} + \frac{5}{9} = -\frac{5 \times 3}{12 \times 3} + \frac{5 \times 4}{9 \times 4}$$

$$-\frac{15}{36} + \frac{20}{36} = \frac{5}{36}$$

Common denominators are not used in multiplication and division. To multiply two fractions, multiply the numerators together and the denominators together. Then, write the result in lowest terms. For example,

$$\frac{2}{3} \times \frac{9}{4} = \frac{18}{12} = \frac{3}{2}$$

Alternatively, the fractions could be factored first to cancel out any common factors before performing the multiplication. For example,

$$\frac{2}{3} \times \frac{9}{4} = \frac{2}{3} \times \frac{3 \times 3}{2 \times 2} = \frac{3}{2}$$

This second approach is helpful when working with larger numbers, as common factors might not be obvious. Multiplication and division of fractions are related because the division of two fractions is changed into a multiplication problem. This means that dividing a fraction by another fraction is the same as multiplying the first fraction by the reciprocal of the second fraction, so that second fraction must be inverted, or "flipped," to be in reciprocal form. For example:

$$\frac{11}{15} \div \frac{3}{5} = \frac{11}{15} \times \frac{5}{3} = \frac{55}{45} = \frac{11}{9}$$

The fraction $\frac{5}{3}$ is the reciprocal of $\frac{3}{5}$.

It is possible to multiply and divide numbers containing a mix of integers and fractions. In this case, convert the integer to a fraction by placing it over a denominator of 1. For example, a division problem involving an integer and a fraction is

$$3 \div \frac{1}{2} = \frac{3}{1} \times \frac{2}{1} = \frac{6}{1} = 6$$

Finally, when performing operations with rational numbers that are negative, the same rules apply as when performing operations with integers. For example, a negative fraction multiplied by a negative fraction results in a positive value, and a negative fraction subtracted from a negative fraction results in a negative value.

Operations can be performed on rational numbers in decimal form. Recall that to write a fraction as an equivalent decimal expression, divide the numerator by the denominator. For example,

$$\frac{1}{8} = 1 \div 8 = 0.125$$

With the case of decimals, it is important to keep track of place value. To add decimals, make sure the decimal places are in alignment so that the numbers are lined up with their decimal points and add vertically. If the numbers do not line up because there are extra or missing place values in one of the numbers, then zeros may be used as placeholders. For example, $0.123 + 0.23$ becomes:

$$
\begin{array}{r}
0.123 \\
+\ 0.230 \\
\hline
0.353
\end{array}
$$

Subtraction is done the same way. Multiplication and division are more complicated. To multiply two decimals, place one on top of the other as in a regular multiplication process and do not worry about lining up the decimal points. Then, multiply as with whole numbers, ignoring the decimals. Finally, in the solution, insert the decimal point as many places to the left as there are total decimal values in the original problem. Here is an example of a decimal multiplication problem:

$$
\begin{array}{r}
0.52 \quad \textit{2 decimal places} \\
\times \ \ 0.2 \quad \textit{1 decimal place} \\
\hline
0.104 \quad \textit{3 decimal places}
\end{array}
$$

The answer to 52 times 2 is 104, and because there are three decimal values in the problem, the decimal point is positioned three units to the left in the answer.

The decimal point plays an integral role throughout the whole problem when dividing with decimals. First, set up the problem in a long division format. If the divisor is not an integer, the decimal must be moved to the right as many units as needed to make it an integer. The decimal in the dividend must be moved to the right the same number of places to maintain equality.

Then, division is completed normally. Here is an example of long division with decimals:

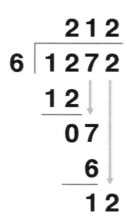

Long division with decimals

Because the decimal point is moved two units to the right in the divisor of 0.06 to turn it into the integer 6, it is also moved two units to the right in the dividend of 12.72 to make it 1,272. The result is 212, and remember that a division problem can always be checked by multiplying the answer by the divisor to see if the result is equal to the dividend.

Sometimes it is helpful to round answers that are in decimal form. First, find the place to which the rounding needs to be done. Then, look at the digit to the right of it. If that digit is 4 or less, the number in the place value to its left stays the same, and everything to its right becomes a 0. This process is known as rounding down. If that digit is 5 or higher, round up by increasing the place value to its left by 1, and every number to its right becomes a 0. If those 0's are in decimals, they can be dropped. For example, 0.145 rounded to the nearest hundredth place would be rounded up to 0.15, and 0.145 rounded to the nearest tenth place would be rounded down to 0.1.

Another operation that can be performed on rational numbers is the square root. Dealing with real numbers only, the **positive square root** of a number is equal to one of the two repeated positive factors of that number. For example

$$\sqrt{49} = \sqrt{7 \cdot 7} = 7$$

A **perfect square** is a number that has a whole number as its square root. Examples of perfect squares are 1, 4, 9, 16, 25, etc. If a number is not a perfect square, an approximation can be used with a calculator. For example

$$\sqrt{67} = 8.185$$

rounded to the nearest thousandth place. The square root of a fraction involving perfect squares involves breaking up the problem into the square root of the numerator separate from the square root of the denominator. For example:

$$\sqrt{\frac{16}{25}} = \frac{\sqrt{16}}{\sqrt{25}} = \frac{4}{5}$$

If the fraction does not contain perfect squares, a calculator can be used. Therefore

$$\sqrt{\frac{2}{5}} = 0.632$$

rounded to the nearest thousandth place. A common application of square roots involves the Pythagorean theorem. Given a right triangle, the sum of the squares of the two legs equals the square of the hypotenuse.

For example, consider the following right triangle:

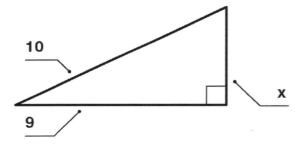

The missing side, x, can be found using the Pythagorean theorem.

$$9^2 + x^2 = 10^2$$

$$81 + x^2 = 100$$

$$x^2 = 19$$

To solve for x, take the square root of both sides. Therefore, $x = \sqrt{19} = 4.36$, which has been rounded to two decimal places.

In addition to the square root, the cube root is another operation. If a number is a **perfect cube**, the cube root of that number is equal to one of the three repeated factors. For example:

$$\sqrt[3]{27} = \sqrt[3]{3 \cdot 3 \cdot 3} = 3$$

Also, unlike square roots, a negative number has a cube root. The result is a negative number. For example:

$$\sqrt[3]{-27} = \sqrt[3]{(-3)(-3)(-3)} = -3$$

Similar to square roots, if the number is not a perfect cube, a calculator can be used to find an approximation. Therefore,

$$\sqrt[3]{\frac{2}{3}} = 0.873$$

rounded to the nearest thousandth place.

Higher-order roots also exist. The number relating to the root is known as the **index.** Given the following root, $\sqrt[3]{64}$, 3 is the index, and 64 is the **radicand.** The entire expression is known as the **radical.** Higher-order roots exist when the index is larger than 3. They can be broken up into two groups: even and odd roots. **Even roots**, when the index is an even number, follow the properties of square roots. A negative number does not have an even root, and an even root is found by finding the single factor that is repeated the same number of times as the index in the radicand.

For example, the fifth root of 32 is equal to 2 because:

$$\sqrt[5]{32} = \sqrt[5]{2 \cdot 2 \cdot 2 \cdot 2 \cdot 2} = 2$$

Odd roots, when the index is an odd number, follow the properties of cube roots. A negative number has an odd root. Similarly, an odd root is found by finding the single factor that is repeated that many times to obtain the radicand. For example, the 4th root of 81 is equal to 3 because $3^4 = 81$. This radical is written as $\sqrt[4]{81} = 3$. Higher-order roots can also be evaluated on fractions and decimals, for example, because:

$$\left(\frac{2}{7}\right)^4 = \frac{16}{2,401}, \sqrt[4]{\frac{16}{2,401}} = \frac{2}{7}$$

and because:

$$(0.1)^5 = 0.00001, \sqrt[5]{0.00001} = 0.1$$

Sometimes, when performing operations in rational numbers, it might be helpful to round the numbers in the original problem to get a rough estimate of what the answer should be.

For example, if you walked into a grocery store and had a $20 bill, your approach might be to round each item to the nearest dollar and add up all the items to make sure that you will have enough money when you check out. This process involves obtaining an estimation of what the exact total would be. In other situations, it might be helpful to round to the nearest $10 amount or $100 amount.

Front-end rounding might be helpful as well in many situations. In this type of rounding, each number is rounded to the highest possible place value. Therefore, all digits except the first digit become 0.

Consider a situation in which you are at the furniture store and want to estimate your total on three pieces of furniture that cost $434.99, $678.99, and $129.99. Front-end rounding would round these three amounts to $400, $700, and $100.

Therefore, the estimate of your total would be $400 + $700 + $100 = $1,200, compared to the exact total of $1,243.97. In this situation, the estimate is not that far off the exact answer. Rounding is useful in both approximating an answer when an exact answer is not needed and for comparison when an exact answer is needed.

For instance, if you had a complicated set of operations to complete and your estimate was $1,000, if you obtained an calculated answer of $100,000, something is off. You might want to check your work to see if a mistake was made because an estimate should not be that different from an exact answer. Estimates can also be helpful with square roots. If a square root of a number is not known, the closest perfect square can be found for an approximation. For example, $\sqrt{50}$ is not equal to a whole number, but 50 is close to 49, which is a perfect square, and $\sqrt{49} = 7$. Therefore, $\sqrt{50}$ is a little bit larger than 7. The actual approximation, rounded to the nearest thousandth, is 7.071.

Ordering and Comparing Rational Numbers

Ordering rational numbers is a way to compare two or more different numerical values. Determining whether two amounts are equal, or if one is less than or greater than the other is the basis for comparing both positive and negative numbers. Also, a group of numbers can be compared by ordering them from the smallest value to the largest value. A few symbols are necessary to use when ordering rational numbers. The equals sign, $=$, shows that the two quantities on either side of the symbol have the same value. For example, $\frac{12}{3} = 4$ because both values are equivalent. Another symbol that is used to compare numbers is $<$, which represents "less than." With this symbol, the smaller number is placed on the left and the larger number is placed on the right. Always remember that the symbol's "mouth" opens up to the larger number. When comparing negative and positive numbers, it is important to remember that the number occurring to the left on the number line is always smaller and is placed to the left of the symbol. This idea might seem confusing because some values could appear at first glance to be larger, even though they are not. For example, $-5 < 4$ is read "negative 5 is less than 4." Here is an image of a number line to help visualize relationships:

The symbol $\leq$ represents "less than or equal to," and it joins $<$ with equality. Therefore, both $-5 \leq 4$ and $-5 \leq -5$ are true statements and "-5 is less than or equal to both 4 and -5." Other symbols are $>$ and $\geq$, which represent "greater than" and "greater than or equal to." Both $4 \geq -1$ and $-1 \geq -1$ are correct ways to use these symbols.

Here is a chart of these four inequality symbols:

Symbol	Definition
$<$	less than
$\leq$	less than or equal to
$>$	greater than
$\geq$	greater than or equal to

Comparing integers is a straightforward process, especially when using the number line, but the comparison of decimals and fractions is not as obvious. When comparing two non-negative decimals, compare digit by digit, starting from the left. The larger value contains the first larger digit. For example, 0.1456 is larger than 0.1234 because the value 4 in the hundredths place in the first decimal is larger than the value 2 in the hundredths place in the second decimal. When comparing a fraction with a decimal, convert the fraction to a decimal and then compare in the same manner.

Finally, there are a few options when comparing multiple fractions. If two non-negative fractions have the same denominator, the fraction with the larger numerator is the larger value. If they have the same numerator but different denominators (such as $\frac{3}{5}$ and $\frac{3}{10}$), the fraction whose denominator is a smaller number (in this case, the $\frac{3}{5}$ because 5 is less than 10) is larger. If the two fractions have different numerators and denominators, they can be converted to equivalent fractions with a common denominator to be compared, or they can be converted to decimals to be compared.

When comparing two negative decimals or fractions, a different approach must be used. It is important to remember that the smaller number exists to the left on the number line. Therefore, when comparing two negative decimals by place value, the number with the larger first place value is smaller due to the negative sign. Whichever value is closer to 0 is larger. For instance, -0.456 is larger than -0.498 because of the values in the hundredth places. If two negative fractions have the same denominator, the fraction with the larger numerator is smaller because of the negative sign.

Solving Real-World One- or Multi-Step Problems with Rational Numbers

One-step problems take only one mathematical step to solve. For example, solving the equation $5x = 45$ is a one-step problem because the one step of dividing both sides of the equation by 5 is the only step necessary to obtain the solution $x = 9$. The **multiplication principle of equality** is the one step used to isolate the variable. The equation is of the form $ax = b$, where a and b are rational numbers. Similarly, the **addition principle of equality** could be the one step needed to solve a problem. In this case, the equation would be of the form $x + a = b$ or $x - a = b$, for real numbers a and b.

A multi-step problem requires more than one step to find the solution, or it might consist of solving more than one equation. An equation that involves both the addition principle and the multiplication principle is a two-step problem, and an example of such an equation is

$$2x - 4 = 5$$

Solving involves adding 4 to both sides and then dividing both sides by 2. An example of a two-step problem involving two separate equations is

$$y = 3x, \ 2x + y = 4$$

The two equations form a system of two equations that must be solved together in two variables. The system can be solved by the substitution method. Since y is already solved for in terms of x, plug $3x$ in for y into the equation

$$2x + y = 4$$

resulting in

$$2x + 3x = 4$$

Therefore, $5x = 4$ and $x = \frac{4}{5}$.

Because there are two variables, the solution consists of both a value for x and for y.

Substitute $x = \frac{4}{5}$ into either original equation to find y.

The easiest choice is $y = 3x$. Therefore:

$$y = 3 \times \frac{4}{5} = \frac{12}{5}$$

The solution can be written as the ordered pair $\left(\frac{4}{5}, \frac{12}{5}\right)$.

Real-world problems can be translated into both one-step and multi-step problems. In either case, the word problem must be translated from the verbal form into mathematical expressions and equations that can be solved using algebra.

An example of a one-step real-world problem is the following: A cat weighs half as much as a dog living in the same house. If the dog weighs 14.5 pounds, how much does the cat weigh? To solve this problem, an equation can be used. In any word problem, the first step is to define variables that represent the unknown quantities. For this problem, let x be equal to the unknown weight of the cat. Because two times the weight of the cat equals 14.5 pounds, the equation to be solved is: $2x = 14.5$. Use the multiplication principle to divide both sides by 2.

Therefore, $x = 7.25$. The cat weighs 7.25 pounds.

Most of the time, real-world problems are more difficult than this one and consist of multi-step problems. The following is an example of a multi-step problem: The sum of two consecutive page numbers is equal to 437. What are those page numbers? First, define the unknown quantities. If x is equal to the first page number, then $x + 1$ is equal to the next page number because they are consecutive integers. Their sum is equal to 437, and this statement translates to the equation:

$$x + x + 1 = 437$$

To solve, first collect like terms to obtain:

$$2x + 1 = 437$$

Then, subtract 1 from both sides and then divide by 2. The solution to the equation is $x = 218$.

Therefore, the two consecutive page numbers that satisfy the problem are 218 and 219. It is always important to make sure that answers to real-world problems make sense. For instance, it should be a red flag if the solution to this same problem resulted in decimals, which would indicate the need to check the work. Page numbers are whole numbers; therefore, if decimals are found to be answers, the solution process should be double-checked to see where mistakes were made.

Factorization

Factorization is the process of breaking up a mathematical quantity, such as a number or polynomial, into a product of two or more factors. For example, a factorization of the number 16 is $16 = 8 \times 2$. If multiplied out, the factorization results in the original number. A **prime factorization** is a specific factorization when the number is factored completely using prime numbers only. For example, the prime factorization of 16 is:

$$16 = 2 \times 2 \times 2 \times 2$$

A factor tree can be used to find the prime factorization of any number.

Within a factor tree, pairs of factors are found until no other factors can be used, as in the following factor tree of number 84:

A factor tree

$$84 = 2 \times 2 \times 3 \times 7$$

It first breaks 84 into 21×4, which is not a prime factorization. Then, both 21 and 4 are factored into their primes. The final numbers on each branch consist of the numbers within the prime factorization. Therefore:

$$84 = 2 \times 2 \times 3 \times 7$$

Factorization can be helpful in finding greatest common divisors and least common denominators.

Also, a factorization of an algebraic expression can be found. Throughout the process, a more complicated expression can be decomposed into products of simpler expressions. To factor a polynomial, first determine if there is a greatest common factor. If there is, factor it out.

For example, $2x^2 + 8x$ has a greatest common factor of $2x$ and can be written as $2x(x + 4)$. Once the greatest common monomial factor is factored out, if applicable, count the number of terms in the polynomial. If there are two terms, is it a difference of squares, a sum of cubes, or a difference of cubes?

If so, the following rules can be used:

$$a^2 - b^2 = (a + b)(a - b)$$

$$a^3 + b^3 = (a + b)(a^2 - ab + b^2)$$

$$a^3 - b^3 = (a - b)(a^2 + ab + b^2)$$

If there are three terms, and if the trinomial is a perfect square trinomial, it can be factored into the following:

$$a^2 + 2ab + b^2 = (a + b)^2$$

$$a^2 - 2ab + b^2 = (a - b)^2$$

If not, try factoring into a product of two binomials by trial and error into a form of $(x + p)(x + q)$. For example, to factor:

$$x^2 + 6x + 8$$

determine what two numbers have a product of 8 and a sum of 6. Those numbers are 4 and 2, so the trinomial factors into:

$$(x + 2)(x + 4)$$

Finally, if there are four terms, try factoring by grouping. First, group terms together that have a common monomial factor. Then, factor out the common monomial factor from the first two terms. Next, look to see if a common factor can be factored out of the second set of two terms that results in a common binomial factor. Finally, factor out the common binomial factor of each expression, for example:

$$xy - x + 5y - 5$$

$$x(y - 1) + 5(y - 1)$$

$$(y - 1)(x + 5)$$

After the expression is completely factored, check to see if the factorization is correct by multiplying to try to obtain the original expression. Factorizations are helpful in solving equations that consist of a polynomial set equal to zero. If the product of two algebraic expressions equals zero, then at least one of the factors is equal to zero. Therefore, factor the polynomial within the equation, set each factor equal to zero, and solve. For example:

$$x^2 + 7x - 18 = 0$$

can be solved by factoring into:

$$(x + 9)(x - 2) = 0$$

Set each factor equal to zero, and solve to obtain $x = -9$ and $x = 2$.

Identifying Integers

Integers include zero, and both positive and negative numbers with no fractional component. Examples of integers are -3, 5, 120, -47, and 0. Numbers that are not integers include 1.3333, ½, -5.7, and 4 ½. Integers can be used to describe different real-world situations. If a scuba diver were to dive 50 feet down into the ocean, his position can be described as -50, in relation to sea level. If, while traveling in Denver, Colorado, a car has an elevation reading of 2300 feet, the integer 2300 can be used to describe the feet above sea level. Integers can be used in many different ways to describe situations with whole numbers and zero.

Integers can also be added and subtracted as situations change. If the temperature in the morning is 45 degrees, and it dropped to 33 degrees in the afternoon, the difference can be found by subtracting the integers 45 and 33 to get a change of 12 degrees. If a submarine was at a depth of 100 feet below sea level, then rose 35 feet, the new depth can be found by adding -100 to 35. The following equation can be used to model the situation with integers:

$$-100 + 35 = -65$$

The answer of -65 reveals the new depth of the submarine, 65 feet below sea level.

Recognizing Rational Exponents

A **rational number** is any number that can be written as a fraction of two integers. Examples of rational numbers include $\frac{1}{2}, \frac{5}{4}$, and 8. The number 8 is rational because it can be expressed as a fraction: $\frac{8}{1} = 8$.

Rational exponents are used to express the root of a number raised to a specific power. For example, $3^{\frac{1}{2}}$ has a base of 3 and rational exponent of $\frac{1}{2}$. The square root of 3 raised to the first power can be written as $\sqrt[2]{3^1}$. Any number with a rational exponent can be written this way. The **numerator**, or number on top of the fraction, becomes the whole number exponent and the **denominator**, or bottom number of the fraction, becomes the root. Another example is $4^{\frac{3}{2}}$. It can be rewritten as the square root of four to the third power, or $\sqrt[2]{4^3}$. This can be simplified by performing the operations 4 to the third power:

$$4^3 = 4 \times 4 \times 4 = 64$$

and then taking the square root of 64, $\sqrt[2]{64}$, which yields an answer of 8. Another way of stating the answer would be 4 to power of $\frac{3}{2}$ is eight, or that 4 to the power of $\frac{3}{2}$ is the square root of 4 cubed:

$$\sqrt[2]{4}^3 = 2^3 = 2 \times 2 \times 2 = 8$$

Understanding Vectors

A **vector** is something that has both magnitude and direction. A vector may sometimes be represented by a ray that has a length, for its magnitude, and a direction. As the magnitude of the vector increases, the length of the ray changes. The direction of the ray refers to the way that the magnitude is applied. The following image shows the placement and parts of a vector.

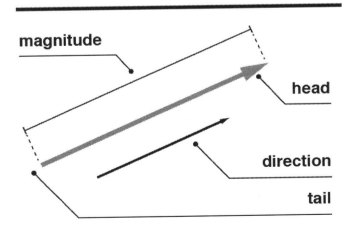

Parts of a Vector

magnitude

head

direction

tail

Examples of vector quantities include force and velocity. **Force** is a vector quantity because applying force requires magnitude, which is the amount of force, and a direction that the force is applied. **Velocity** is a vector because it has a magnitude, or speed that an object travels, and also the direction that the object is traveling in. Vectors can be added together by placing the tail of the second at the head of the first. The resulting vector is found by starting at the first tail and ending at the second head. An example of this is show in the following picture.

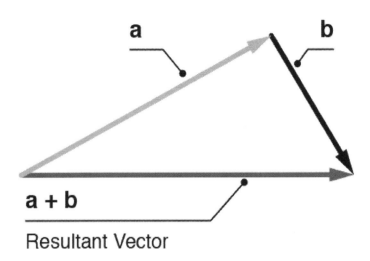

Subtraction can also be done with vectors by adding the inverse of the second vector. The inverse is found by reversing the direction of the vector. Then addition can take place just as described above, but using the inverse instead of the original vector. Scalar multiplication can also be done with vectors. This multiplication changes the magnitude of the vector by the **scalar,** or number. For example, if the length is described as 4, then scalar multiplication is used to multiply by 2, so the vector magnitude becomes 8. The direction of the vector is left unchanged because scalar does not include direction.

Vectors may also be described using coordinates on a plane, such as (5, 2). This vector would start at a point and move to the right 5 and up 2. The two coordinates describe the horizontal and vertical components of the vector. The starting point in relation to the coordinates is the tail, and the ending point is the head.

Creating Matrices
A **matrix** is an arrangement of numbers in rows and columns. Matrices are used to work with vectors and transform them. One example is a system of linear equations. Matrices can represent a system and be used to transform and solve the system. An important connection between scalars, vectors, and matrices is this: scalars are only numbers, vectors are numbers with magnitude and direction, and matrices are an array of numbers in rows and columns. The rows run from left to right and the columns run from top to bottom. When describing the dimensions of a matrix, the number of rows is stated first and the number of columns is stated second. The following matrix has two rows and three columns, referred to as a 2×3 matrix:

$$\begin{bmatrix} 3 & 5 & 7 \\ 4 & 2 & 8 \end{bmatrix}$$

A number in a matrix can be found by describing its location. For example, the number in row two, column three is 8. In row one, column two, the number 5 is found.

Operations can be performed on matrices, just as they can on vectors. Scalar multiplication can be performed on matrices and it will change the magnitude, just as with a vector. A scalar multiplication problem using a 2×2 matrix looks like the following:

$$3 \times \begin{bmatrix} 4 & 5 \\ 8 & 3 \end{bmatrix}$$

The scalar of 3 is multiplied by each number to form the resulting matrix:

$$\begin{bmatrix} 12 & 15 \\ 24 & 9 \end{bmatrix}$$

Matrices can also be added and subtracted. For these operations to be performed, the matrices must be the same dimensions. Other operations that can be performed to manipulate matrices are multiplication, division, and transposition. **Transposing** a matrix means to switch the rows and columns. If the original matrix has two rows and three columns, then the transposed matrix has three rows and two columns.

Algebra

Solving, Graphing, and Modeling Multiple Types of Expressions
When presented with a real-world problem that must be solved, the first step is always to determine what the unknown quantity is that must be solved for. Use a **variable**, such as x or t, to represent that unknown quantity. Sometimes there can be two or more unknown quantities. In this case, either choose an additional variable, or if a relationship exists between the unknown quantities, express the other quantities in terms of the original variable. After choosing the variables, form algebraic expressions and/or equations that represent the verbal statement in the problem. The following table shows examples of vocabulary used to represent the different operations:

Addition	Sum, plus, total, increase, more than, combined, in all
Subtraction	Difference, less than, subtract, reduce, decrease, fewer, remain
Multiplication	Product, multiply, times, part of, twice, triple
Division	Quotient, divide, split, each, equal parts, per, average, shared

The combination of operations and variables form both mathematical expression and equations. The difference between expressions and equations are that there is no equals sign in an expression, and that expressions are **evaluated** to find an unknown quantity, while equations are **solved** to find an unknown quantity. Also, inequalities can exist within verbal mathematical statements. Instead of a statement of equality, expressions state quantities are *less than, less than or equal to, greater than,* or *greater than or equal to.* Another type of inequality is when a quantity is said to be *not equal to* another quantity. The symbol used to represent "not equal to" is ≠.

The steps for solving inequalities in one variable are the same steps for solving equations in one variable. The addition and multiplication principles are used. However, to maintain a true statement when using the $<, \leq, >$, and $\geq$ symbols, if a negative number is either multiplied times both sides of an inequality or divided from both sides of an inequality, the sign must be flipped. For instance, consider the following inequality:

$$3 - 5x \leq 8$$

First, 3 is subtracted from each side to obtain $-5x \leq 5$. Then, both sides are divided by -5, while flipping the sign, to obtain $x \geq -1$. Therefore, any real number greater than or equal to -1 satisfies the original inequality.

Adding and Subtracting Linear Algebraic Expressions

To add and subtract linear algebra expressions, you must combine like terms. **Like terms** are described as those terms that have the same variable with the same exponent. In the following example, the x-terms can be added because the variable is the same and the exponent on the variable of one is also the same. These terms add to be $9x$. The other like terms are called **constants** because they have no variable component. These terms will add to be nine.

Example: Add $(3x - 5) + (6x + 14)$

$3x - 5 + 6x + 14$ Rewrite without parentheses

$3x + 6x - 5 + 14$ Commutative property of addition

$9x + 9$ Combine like terms

When subtracting linear expressions, be careful to add the opposite when combining like terms. Do this by distributing -1, which is multiplying each term inside the second parenthesis by negative one. Remember that distributing -1 changes the sign of each term.

Example: Subtract $(17x + 3) - (27x - 8)$

$17x + 3 - 27x + 8$ Distributive Property

$17x - 27x + 3 + 8$ Commutative property of addition

$-10x + 11$ Combine like terms

Example: Simplify by adding or subtracting:

$(6m + 28z - 9) + (14m + 13) - (-4z + 8m + 12)$

$6m + 28z - 9 + 14m + 13 + 4z - 8m - 12$ Distributive Property

$6m + 14m - 8m + 28z + 4z - 9 + 13 - 12$ Commutative Property of Addition

$12m + 32z - 8$ Combine like terms

Solving Problems Using Numerical and Algebraic Expressions

Translating sentences describing relationships between variables and constants to algebraic expressions and equations involves recognizing key words that represent mathematical operations. This process is known as **modeling.** For simplicity, let x be the variable, or the unknown quantity. Statements that include the four operations addition, subtraction, multiplication, and division exist in sentences that model linear relationships.

For example, words and phrases that represent addition are "sum," "more than," and "increased by." Words and phrases that represent subtraction are "minus," "decreased by," "subtracted from," "difference," "less", "fewer than," and "less than." Words and phrases that represent multiplication are "times," "product of," "twice," "double," and "triple." Finally, words and phrases that represent division are "divided by," "quotient," and "reciprocal." Most of the time, these words and phrases are combined to represent expressions that deal with one or more operation. For example, "ten subtracted from nine times a number" would be represented as $9x - 10$, and "the quotient of a number and 7 increased by 8" would be represented as $\frac{x}{7} + 8$. The word problems that typically use these expressions will have a statement of equality. For instance, the problem could say "ten subtracted from nine times a number equals 20; find that number." In this case, the algebraic expression shown previously would be set equal to 20 and then solved for x.

$$9x - 10 = 20$$

Add ten to both sides and then divide by 9 to get the solution $x = \frac{30}{9}$, which reduces to $x = \frac{10}{3}$.

Other types of expressions, besides linear expressions, can be the results of modeling. If the variable is raised to a power other than 1, the result is a **polynomial expression**. The path of an object thrown up into the air is a common example of this. The graph of an object represents an upside-down parabola, which is modeled by an equation of the type $y = -ax^2$. In this case, a represents the height of the object at its highest point before coming back down to the ground, and the negative sign shows that the parabola is upside down.

Here is the graph of a parabola:

Parabola

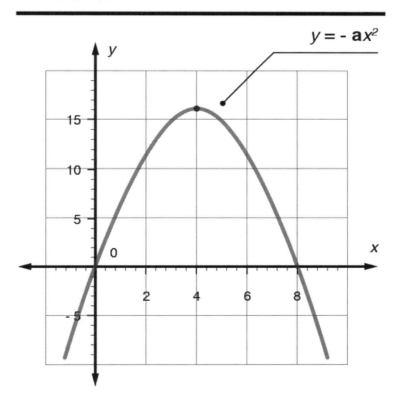

Evaluating and Simplifying Algebraic Expressions

Given an algebraic expression, students may be asked to evaluate for given values of variable(s). In doing so, students will arrive at a numerical value as an answer. For example:

$$\text{Evaluate } a - 2b + ab \text{ } for \text{ } a = 3 \text{ and } b = -1$$

To evaluate an expression, the given values should be substituted for the variables and simplified using the order of operations. In this case: $(3) - 2(-1) + (3)(-1)$. Parentheses are used when substituting.

Given an algebraic expression, students may be asked to simplify the expression. For example:

$$\text{Simplify } 5x^2 - 10x + 2 - 8x^2 + x - 1.$$

Simplifying algebraic expressions requires combining like terms. A term is a number, variable, or product of a number and variables separated by addition and subtraction. The terms in the above expression are: $5x^2, -10x, 2, -8x^2, x$, and -1. Like terms have the same variables raised to the same powers (exponents). To combine like terms, the coefficients (numerical factor of the term including sign) are added, while the variables and their powers are kept the same. The example above simplifies to:

$$-3x^2 - 9x + 1$$

Generating Equivalent Expressions

Two algebraic expressions are equivalent if, even though they look different, they represent the same expression. Therefore, plugging in the same values into the variables in each expression will result in the same result in both expressions. To obtain an equivalent form of an algebraic expression, laws of algebra must be followed. For instance, addition and multiplication are both commutative and associative. Therefore, terms in an algebraic expression can be added in any order and multiplied in any order.

For instance, $4x + 2y$ is equivalent to $2y + 4x$ and $y \times 2 + x \times 4$.

Also, the distributive law allows a number to be distributed throughout parentheses, as in the following:

$$a(b + c) = ab + ac$$

The two expressions on both sides of the equals sign are equivalent. Also, collecting like terms is important when working with equivalent forms. The simplest version of an expression is always the one easiest to work with, so all like terms (those with the same variables raised to the same powers) must be combined.

Note that an expression is not an equation; therefore, expressions cannot be multiplied by numbers, divided by numbers, or have numbers added to them or subtracted from them and still have equivalent expressions. These processes can only happen in equations when the same step is performed on both sides of the equals sign.

The **distributive property** ($a(b + c) = ab + ac$) is a way of taking a factor and multiplying it through a given expression in parentheses. Each term inside the parentheses is multiplied by the outside factor, eliminating the parentheses. The following example shows how to distribute the number 3 to all the terms inside the parentheses.

Example: Use the distributive property to write an equivalent algebraic expression:

$$3(2x + 7y + 6)$$

$$3(2x) + 3(7y) + 3(6) \qquad \text{Distributive property}$$

$$6x + 21y + 18 \qquad \text{Simplify}$$

Because $a - b$ can be written $a + (-b)$, the distributive property can be applied in the example below:

Example: Use the distributive property to write an equivalent algebraic expression.

$$7(5m - 8)$$

$$7[5m + (-8)] \qquad \text{Rewrite subtraction as addition of } -8$$

$$7(5m) + 7(-8) \qquad \text{Distributive property}$$

$$35m - 56 \qquad \text{Simplify}$$

In the following example, note that the factor of 2 is written to the right of the parentheses but is still distributed as before.

Example: Use the distributive property to write an equivalent algebraic expression:

$$(3m + 4x - 10)2$$

$$(3m)2 + (4x)2 + (-10)2 \qquad \text{Distributive property}$$

$$6m + 8x - 20 \qquad \text{Simplify}$$

Example: $\quad -(-2m + 6x)$

In this example, the negative sign in front of the parentheses can be interpreted as $-1(-2m + 6x)$

$$-1(-2m + 6x)$$

$$-1(-2m) + (-1)(6x) \qquad \text{Distributive property}$$

$$2m - 6x \qquad \text{Simplify}$$

Linear Equations

An **equation in one variable** is a mathematical statement where two algebraic expressions with one variable, usually x, are set equal. To solve the equation, the variable must be isolated on one side of the equals sign. The addition and multiplication principles of equality are used to isolate the variable. The **addition principle of equality** states that the same number can be added to or subtracted from both sides of an equation. Because the same value is being used on both sides of the equals sign, equality is maintained.

For example, the equation $2x = 5x$ is equivalent to both $2x + 3 = 5x + 3$, and $2x - 5 = 5x - 5$.

This principle can be used to solve the following equation: $x + 5 = 4$.

The variable x must be isolated, so to move the 5 from the left side, subtract 5 from both sides of the equals sign.

Therefore:

$$x + 5 - 5 = 4 - 5$$

So, the solution is $x = -1$.

This process illustrates the idea of an **additive inverse** because subtracting 5 is the same as adding -5. Basically, add the opposite of the number that must be removed to both sides of the equals sign. The **multiplication principle of equality** states that equality is maintained when a number is either multiplied times both expressions on each side of the equals sign, or when both expressions are divided by the same number.

For example, $4x = 5$ is equivalent to both $16x = 20$ and $x = \frac{5}{4}$.

Multiplying both sides by 4 and dividing both sides by 4 maintains equality. Solving the equation:

$$6x - 18 = 5$$

requires the use of both principles.

First, apply the addition principle to add 18 to both sides of the equals sign, which results in $6x = 23$.

Then use the multiplication principle to divide both sides by 6, giving the solution $x = \frac{23}{6}$.

Using the multiplication principle in the solving process is the same as involving a multiplicative inverse. A **multiplicative inverse** is a value that, when multiplied by a given number, results in 1. Dividing by 6 is the same as multiplying by $\frac{1}{6}$, which is both the reciprocal and multiplicative inverse of 6.

When solving a linear equation in one variable, checking the answer shows if the solution process was performed correctly. Plug the solution into the variable in the original equation. If the result is a false statement, something was done incorrectly during the solution procedure. Checking the example above gives the following:

$$6 \times \frac{23}{6} - 18 = 23 - 18 = 5$$

Therefore, the solution is correct.

Some equations in one variable involve fractions or the use of the distributive property. In either case, the goal is to obtain only one variable term and then use the addition and multiplication principles to isolate that variable.

Consider the equation $\frac{2}{3}x = 6$.

To solve for x, multiply each side of the equation by the reciprocal of $\frac{2}{3}$, which is $\frac{3}{2}$.

This step results in $\frac{3}{2} \times \frac{2}{3}x = \frac{3}{2} \times 6$, which simplifies into the solution $x = 9$.

Now consider the equation:

$$3(x + 2) - 5x = 4x + 1$$

Use the distributive property to clear the parentheses. Therefore, multiply each term inside the parentheses by 3.

This step results in:

$$3x + 6 - 5x = 4x + 1$$

Next, collect like terms on the left-hand side. **Like terms** are terms with the same variable or variables raised to the same exponent(s). Only like terms can be combined through addition or subtraction.

After collecting like terms, the equation is:

$$-2x + 6 = 4x + 1$$

Finally, apply the addition and multiplication principles. Add $2x$ to both sides to obtain $6 = 6x + 1$.

Then, subtract 1 from both sides to obtain $5 = 6x$. Finally, divide both sides by 6 to obtain the solution:

$$\frac{5}{6} = x$$

Two other types of solutions can be obtained when solving an equation in one variable. The final result could be that there is either no solution or that the solution set contains all real numbers. Consider the equation:

$$4x = 6x + 5 - 2x$$

First, the like terms can be combined on the right to obtain:

$$4x = 4x + 5$$

Next, subtract $4x$ from both sides. This step results in the false statement $0 = 5$. There is no value that can be plugged into x that will ever make this equation true. Therefore, there is no solution. The solution procedure contained correct steps, but the result of a false statement means that no value satisfies the equation. The symbolic way to denote that no solution exists is $\emptyset$. Next, consider the equation:

$$5x + 4 + 2x = 9 + 7x - 5$$

Combining the like terms on both sides results in:

$$7x + 4 = 7x + 4$$

The left-hand side is exactly the same as the right-hand side. Using the addition principle to move terms, the result is $0 = 0$, which is always true. Therefore, the original equation is true for any number, and the solution set is all real numbers. The symbolic way to denote such a solution set is $\mathbb{R}$, or in interval notation, $(-\infty, \infty)$.

Algebraically Solving Linear Equations or Inequalities in One Variable
A *linear equation in one variable* can be solved using the following steps:

1. Simplify the algebraic expressions on both sides of the equals sign by removing all parentheses, using the distributive property, and then collect all like terms.

2. Collect all variable terms on one side of the equals sign and all constant terms on the other side by adding the same quantity to both sides of the equals sign, or by subtracting the same quantity from both sides of the equals sign.

3. Isolate the variable by either dividing both sides of the equation by the same number, or by multiplying both sides by the same number.

4. Check the answer.

The only difference between solving linear inequalities versus equations is that when multiplying by a negative number or dividing by a negative number, the direction of the inequality symbol must be reversed.

If an equation contains multiple fractions, it might make sense to clear the equation of fractions first by multiplying all terms by the least common denominator. Also, if an equation contains several decimals, it might make sense to clear the decimals as well by multiplying times a factor of 10. If the equation has decimals in the hundredths place, multiply every term in the equation by 100.

Polynomial Equations
A **polynomial** is a mathematical expression containing the sum and difference of one or more terms that are constants multiplied times variables raised to positive powers. A polynomial is considered expanded when there are no variables contained within parentheses, the distributive property has been carried out for any terms that were within parentheses, and like terms have been collected.

When working with polynomials, **like terms** are terms that contain exactly the same variables with the same powers. For example, x^4y^5 and $9x^4y^5$ are like terms. The coefficients are different, but the same variables are raised to the same powers. When adding polynomials, only terms that are considered like terms can be added. When adding two like terms, just add the coefficients and leave the variables alone. This process uses the distributive property. For example:

$$x^4y^5 + 9x^4y^5$$
$$(1+9)x^4y^5$$
$$10x^4y^5$$

Therefore, when adding two polynomials, simply add the like terms together. Unlike terms cannot be combined.

Subtracting polynomials involves adding the opposite of the polynomial being subtracted. Basically, the sign of each term in the polynomial being subtracted is changed, and then the like terms are combined because it is now an addition problem. For example, consider the following:

$$6x^2 - 4x + 2 - (4x^2 - 8x + 1)$$

Add the opposite of the second polynomial to obtain:

$$6x^2 - 4x + 2 + (-4x^2 + 8x - 1)$$

Then, collect like terms to obtain:

$$2x^2 + 4x + 1$$

Multiplying polynomials involves using the product rule for exponents that $b^m b^n = b^{m+n}$. Basically, when multiplying expressions with the same base, just add the exponents. Multiplying a monomial by a monomial involves multiplying the coefficients together and then multiplying the variables together using the product rule for exponents. For instance:

$$8x^2y \times 4x^4y^2 = 32x^6y^3$$

When multiplying a monomial by a polynomial that is not a monomial, use the distributive property to multiply each term of the polynomial times the monomial. For example:

$$3x(x^2 + 3x - 4) = 3x^3 + 9x^2 - 12x$$

Finally, multiplying two polynomials when neither one is a monomial involves multiplying each term of the first polynomial by each term of the second polynomial. There are some shortcuts, given certain scenarios. For instance, a binomial times a binomial can be found by using the **FOIL (Firsts, Outers, Inners, Lasts)** method shown here.

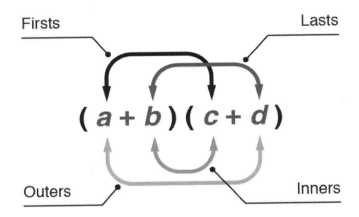

Finding the product of a sum and difference of the same two terms is simple because if it was to be foiled out, the outer and inner terms would cancel out. For instance:

$$(x + y)(x - y) = x^2 + xy - xy - y^2$$

Finally, the square of a binomial can be found using the following formula:

$$(a \pm b)^2 = a^2 \pm 2ab + b^2$$

Radical and Exponential Relationships

The *n*th root of a is given as $\sqrt[n]{a}$, which is called a **radical.** Typical values for *n* are 2 and 3, which represent the square and cube roots. In this form, *n* represents an integer greater than or equal to 2, and *a* is a real number. If *n* is even, *a* must be nonnegative, and if *n* is odd, *a* can be any real number. This radical can be written in exponential form as $a^{\frac{1}{n}}$. Therefore, $\sqrt[4]{15}$ is the same as $15^{\frac{1}{4}}$ and $\sqrt[3]{-5}$ is the same as $(-5)^{\frac{1}{3}}$.

In a similar fashion, the nth root of a can be raised to a power m, which is written as $\left(\sqrt[n]{a}\right)^m$. This expression is the same as $\sqrt[n]{a^m}$. For example:

$$\sqrt[2]{4^3} = \sqrt[2]{64} = 8 = \left(\sqrt[2]{4}\right)^3 = 2^3$$

Because $\sqrt[n]{a} = a^{\frac{1}{n}}$, both sides can be raised to an exponent of m, resulting in:

$$\left(\sqrt[n]{a}\right)^m = \sqrt[n]{a^m} = a^{\frac{m}{n}}$$

This rule allows:

$$\sqrt[2]{4^3} = \left(\sqrt[2]{4}\right)^3 = 4^{\frac{3}{2}}$$

$$(2^2)^{\frac{3}{2}} = 2^{\frac{6}{2}} = 2^3 = 8$$

Negative exponents can also be incorporated into these rules. Any time an exponent is negative, the base expression must be flipped to the other side of the fraction bar and rewritten with a positive exponent. For instance:

$$2^{-3} = \frac{1}{2^3} = \frac{1}{8}$$

Therefore, two more relationships between radical and exponential expressions are:

$$a^{-\frac{1}{n}} = \frac{1}{\sqrt[n]{a}}$$

and

$$a^{-\frac{m}{n}} = \frac{1}{\sqrt[n]{a^m}} = \frac{1}{\left(\sqrt[n]{a}\right)^m}$$

Thus:

$$8^{-\frac{1}{3}} = \frac{1}{\sqrt[3]{8}} = \frac{1}{2}$$

All of these relationships are very useful when simplifying complicated radical and exponential expressions. If an expression contains both forms, use one of these rules to change the expression to contain either all radicals or all exponential expressions. This process makes the entire expression much easier to work with, especially if the expressions are contained within equations.

Consider the following example:

$$\sqrt{x} \times \sqrt[4]{x}$$

It is written in radical form; however, it can be simplified into one radical by using exponential expressions first. The expression can be written as:

$$x^{\frac{1}{2}} \times x^{\frac{1}{4}}$$

It can be combined into one base by adding the exponents as:

$$x^{\frac{1}{2}+\frac{1}{4}} = x^{\frac{3}{4}}$$

Writing this back in radical form, the result is $\sqrt[4]{x^3}$.

Systems of Equations
Creating, Solving, or Interpreting Systems of Linear Inequalities in Two Variables

A **system of linear inequalities in two variables** consists of two inequalities in two variables, typically x and y. For example, the following is a system of linear inequalities in two variables:

$$\begin{cases} 4x + 2y < 1 \\ 2x - y \leq 0 \end{cases}$$

The curly brace on the left side shows that the two inequalities are grouped together. A solution of a single inequality in two variables is an ordered pair that satisfies the inequality. For example, (1, 3) is a solution of the linear inequality $y \geq x + 1$ because when plugged in, it results in a true statement. The graph of an inequality in two variables consists of all ordered pairs that make the solution true. Therefore, the entire solution set of a single inequality contains many ordered pairs, and the set can be graphed by using a half plane. A **half plane** consists of the set of all points on one side of a line. If the inequality consists of $>$ or $<$, the line is dashed because no solutions actually exist on the line shown. If the inequality consists of $\geq$ or $\leq$, the line is solid and solutions are on the line shown. To graph a linear inequality, graph the corresponding equation found by replacing the inequality symbol with an equals sign. Then pick a test point that exists on either side of the line. If that point results in a true statement when plugged into the original inequality, shade in the side containing the test point. If it results in a false statement, shade in the opposite side.

Solving a system of linear inequalities must be done graphically. Follow the process as described above for both given inequalities. The solution set to the entire system is the region that is in common to every graph in the system. For example, here is the solution to the following system:

$$\begin{cases} y \geq 3 - x \\ y \leq -3 - x \end{cases}$$

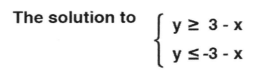

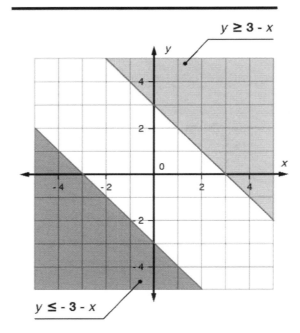

Note that there is no region in common, so this system has no solution.

Creating, Solving, or Interpreting Systems of Two Linear Equations in Two Variables
An example of a system of two linear equations in two variables is the following:

$$2x + 5y = 8$$

$$5x + 48y = 9$$

A solution to a **system of two linear equations** is an ordered pair that satisfies both the equations in the system. A system can have one solution, no solution, or infinitely many solutions. The solution can be found through a graphing technique. The solution of a system of equations is actually equal to the point of intersection of both lines. If the lines intersect at one point, there is one solution and the system is said to be **consistent**. However, if the two lines are parallel, they will never intersect and there is no solution. In this case, the system is said to be **inconsistent**. Thirdly, if the two lines are actually the same line, there are infinitely many solutions and the solution set is equal to the entire line. The lines are **dependent**.

Here is a summary of the three cases:

Solving Systems by Graphing

Consistent	Inconsistent	Dependent
One solution	No solution	Infinite number of solutions
Lines intersect	*Lines are parallel*	*Coincide: same line*

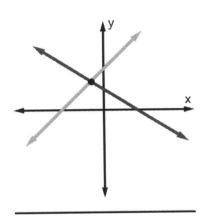 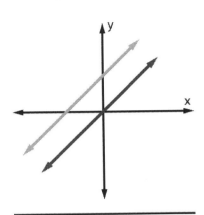 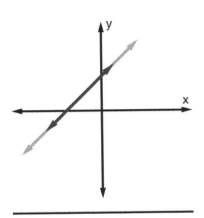

Consider the following system of equations:

$$\begin{cases} y + x = 3 \\ y - x = 1 \end{cases}$$

To find the solution graphically, graph both lines on the same *xy*-plane. Graph each line using either a table of ordered pairs, the *x*- and *y*-intercepts, or slope and the *y*-intercept. Then, locate the point of intersection.

The graph is shown here:

The System of Equations $\begin{cases} y + x = 3 \\ y - x = 1 \end{cases}$

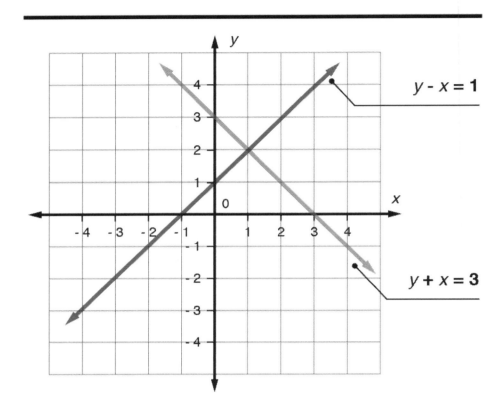

It can be seen that the point of intersection is the ordered pair (1, 2). This solution can be checked by plugging it back into both original equations to make sure it results in true statements. This process results in:

$$2 + 1 = 3$$

$$2 - 1 = 1$$

Both are true equations, so therefore the point of intersection is truly the solution.

The following system has no solution:

$$y = 4x + 1$$

$$y = 4x - 1$$

Both lines have the same slope and different y-intercepts; therefore, they are parallel. This means that they run alongside each other and never intersect.

Finally, the following solution has infinitely many solutions:

$$2x - 7y = 12$$

$$4x - 14y = 24$$

Note that the second equation is equal to the first equation times 2. Therefore, they are the same line. The solution set can be written in set notation as $\{(x, y) | 2x - 7y = 12\}$, which represents the entire line.

Algebraically Solving Systems of Two Linear Equations in Two Variables
There are two algebraic methods to finding solutions. The first is **substitution**. This process is better suited for systems when one of the equations is already solved for one variable, or when solving for one variable is easy to do. The equation that is already solved for is substituted into the other equation for that variable, and this process results in a linear equation in one variable. This equation can be solved for the given variable, and then that solution can be plugged into one of the original equations, which can then be solved for the other variable. This last step is known as **back-substitution** and the end result is an ordered pair.

A system that is best suited for substitution is the following:

$$y = 4x + 2$$

$$2x + 3y = 9$$

The other method is known as **elimination,** or the **addition method**. This is better suited when the equations are in standard form $Ax + By = C$. The goal in this method is to multiply one or both equations times numbers that result in opposite coefficients. Then, add the equations together to obtain an equation in one variable. Solve for the given variable, then take that value and back-substitute to obtain the other part of the ordered pair solution.

A system that is best suited for elimination is the following:

$$2x + 3y = 8$$

$$4x - 2y = 10$$

Note that in order to check an answer when solving a system of equations, the solution must be checked in both original equations to show that it solves not only one of the equations, but both of them.

If throughout either solution procedure the process results in an untrue statement, there is no solution to the system. Finally, if throughout either solution procedure the process results in the variables dropping out, which gives a statement that is always true, there are infinitely many solutions.

Functions

Function Definition

A **relation** is any set of ordered pairs (x, y). The first set of points, known as the x-coordinates, make up the **domain** of the relation. The second set of points, known as the y-coordinates, make up the **range** of the relation. A relation in which every member of the domain corresponds to only one member of the range is known as a **function**. A function cannot have a member of the domain corresponding to two members of the range.

Function Notation

Functions are most often given in terms of equations instead of ordered pairs. For instance, here is an equation of a line:

$$y = 2x + 4$$

In function notation, this can be written as

$$f(x) = 2x + 4$$

The expression $f(x)$ is read "f of x" and it shows that the inputs, the x-values, get plugged into the function and the output is $y = f(x)$. The set of all inputs are in the domain and the set of all outputs are in the range.

The x-values are known as the **independent variables** of the function and the y-values are known as the **dependent variables** of the function. The y-values depend on the x-values. For instance, if x = 2 is plugged into the function shown above, the y-value depends on that input.

$$f(2) = 2 \times 2 + 4 = 8.$$

Therefore, $f(2) = 8$, which is the same as writing the ordered pair (2, 8). To graph a function, graph it in equation form. Therefore, replace $f(x)$ with h and plot ordered pairs.

Due to the definition of a function, the graph of a function cannot have two of the same x-components paired to different y-component. For example, the ordered pairs (3, 4) and (3, -1) cannot be in a valid function. Therefore, all graphs of functions pass the **vertical line test**. If any vertical line intersects a graph in more than one place, the graph is not that of a function. For instance, the graph of a circle is not a function because one can draw a vertical line through a circle and the line would intersect the circle twice. Common functions include lines and polynomials, and they all pass the vertical line test.

Linear Functions

A linear function that models a linear relationship between two quantities is of the form $y = mx + b$, or in function form

$$f(x) = mx + b$$

In a linear function, the value of y depends on the value of x, and y increases or decreases at a constant rate as x increases. Therefore, the independent variable is x, and the dependent variable is y. The graph of a linear function is a line, and the constant rate can be seen by looking at the steepness, or slope, of the line. If the line increases from left to right, the slope is positive. If the line slopes downward from left to right, the slope is negative. In the function, m represents slope. Each point on the line is an **ordered pair** (x, y), where x represents the x-coordinate of the point and y represents the y-coordinate of the point. The point where x = 0 is known as the y-intercept, and it is the place where the line crosses the y-axis. If x = 0 is plugged into $f(x) = mx + b$, the result is $f(0) = b$; therefore, the point (0, b) is the y-intercept of the line. The derivative of a linear function is its slope.

Consider the following situation. A taxicab driver charges a flat fee of $2 per ride and $3 a mile. This statement can be modeled by the function $f(x) = 3x + 2$ where x represents the number of miles and $f(x) = y$ represents the total cost of the ride. The total cost increases at a constant rate of $2 per mile, and that is why this situation is a linear relationship. The slope $m = 3$ is equivalent to this rate of change. The flat fee of $2 is the y-intercept. It is the place where the graph crosses the x-axis, and it represents the cost when $x = 0$, or when no miles have been traveled in the cab. The y-intercept in this situation represents the flat fee.

A linear function of the form $f(x) = mx + b$ has two important quantities: m and b. The quantity m represents the slope of the line, and the quantity b represents the y-intercept of the line. When the function represents an actual real-life situation, or mathematical model, these two quantities are very meaningful. The slope, m, represents the rate of change, or the amount y increases or decreases given an increase in x. If m is positive, the rate of change is positive, and if m is negative, the rate of change is negative. The y-intercept, b, represents the amount of the quantity y when x is 0. In many applications, if the x-variable is never a negative quantity, the y-intercept represents the initial amount of the quantity y. Often the x-variable represents time, so it makes sense that the x-variable is never negative.

Consider the following example. These two equations represent the cost, C, of t-shirts, x, at two different printing companies:

$$C(x) = 7x$$

$$C(x) = 5x + 25$$

The first equation represents a scenario that shows the cost per t-shirt is $7. In this equation, x varies directly with y. There is no y-intercept, which means that there is no initial cost for using that printing company. The rate of change is 7, which is price per shirt. The second equation represents a scenario that has both an initial cost and a cost per t-shirt. The slope 5 shows that each shirt is $5. The y-intercept 25 shows that there is an initial cost of $25 when using that company. Therefore, it makes sense to use the first company at $7 a shirt when only purchasing a small number of t-shirts. However, large orders would be cheaper by going with the second company because eventually that initial cost will become negligible.

Radical Functions
Recall that a **radical expression** is an expression involving a square root, a cube root, or a higher order root such as fourth root, fifth root, etc. The expression underneath the radical is known as the **radicand** and the index is the number corresponding to the **root**. An index of 2 corresponds to a square root. A radical function is a function that involves a radical expression.

For instance, $\sqrt{x + 1}$ is a radical expression, $x + 1$ is the radicand, and the corresponding function is:

$$y = \sqrt{x + 1}$$

The function can also be written in function notation as:

$$f(x) = \sqrt{x + 1}$$

If the root is even, meaning a square root, fourth root, etc., the radicand must be positive. Therefore, in order to find the domain of a radical function with an even index, set the radicand greater than or equal to zero and find the set of numbers that satisfies that inequality. The domain of:

$$f(x) = \sqrt{x + 1}$$

is all numbers greater than or equal to -1. The range of this function is all nonnegative real numbers because the square root, or any even root, can never output a negative number. The domain of an odd root is all real numbers because the radicand can be negative in an odd root.

Piecewise Functions

A **piecewise function** is basically a function that is defined in pieces. The graph of the function behaves differently over different intervals along the x-axis, or different intervals of its domain. Therefore, the function is defined using different mathematical expressions over these intervals. The function is not defined by only one equation. In a piecewise function, the function is actually defined by two or more equations, where each equation is used over a specific interval of the domain.

Here is a graph of a piecewise function:

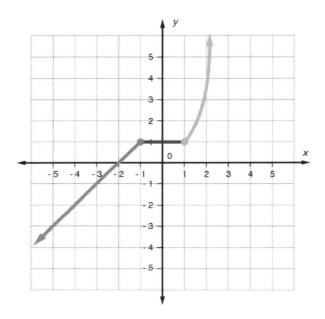

Notice that from $(-\infty, -1]$, the graph is a line with positive slope. From $[-1, 1]$ the graph is a horizontal line. Finally, from $[1, \infty)$ the graph is a nonlinear curve. Both the domain and range of this piecewise defined function is all real numbers, which is expressed as $(-\infty, \infty)$.

Piecewise functions can also have **discontinuities**, which are jumps in the graph. When drawing a graph, if the pencil must be picked up at any point to continue drawing, the graph has a discontinuity. Here is the graph of a piecewise function with discontinuities at $x = 1$ and $x = 2$:

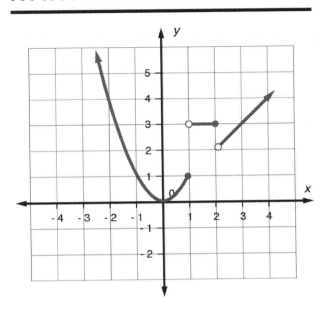

A Piecewise Function

The open circle at a point indicates that the endpoint is not included in that part of the graph, and the closed circle indicates that the endpoint is included. The domain of this function is all real numbers; however, the range is all nonnegative real numbers $[0, \infty)$.

Polynomial Functions
A **polynomial function** is a function containing a polynomial expression, which is an expression containing constants and variables combined using the four mathematical operations. The degree of a polynomial in one variable is the largest exponent seen on any variable in the expression. Typical polynomial functions are **quartic,** with a degree of 4, **cubic,** with a degree of 3, and **quadratic,** with a degree of 2. Note that the exponents on the variables can only be nonnegative integers. The domain of any polynomial function is all real numbers because any number plugged into a polynomial expression grants a real number output.

An example of a quartic polynomial equation is:

$$y = x^4 + 3x^3 - 2x + 1$$

The zeros of a polynomial function are the points where its graph crosses the y-axis. In order to find the number of real zeros of a polynomial function, **Descartes' Rule of Sign** can be used. The number of possible positive real zeros is equal to the number of sign changes in the coefficients of the terms in the polynomial. If there is only one sign change, there is only one positive real zero. In the example above, the signs of the coefficients are positive, positive, negative, and positive.

Therefore, the sign changes two times and therefore, there are at most two positive real zeros. The number of possible negative real zeros is equal to the number of sign changes in the coefficients when plugging $-x$ into the equation. Again, if there is only one sign change, there is only one negative real zero. The polynomial result when plugging $-x$ into the equation is:

$$y^4 - 3x^3 + 2x + 1$$

The sign changes two times, so there are at most two negative real zeros. Another polynomial equation this rule can be applied to is:

$$y = x^3 + 2x - x - 5$$

There is only one sign change in the terms of the polynomial, so there is exactly one real zero. When plugging $-x$ into the equation, the polynomial result is:

$$-x^3 - 2x - x - 5$$

There are no sign changes in this polynomial, so there are no possible negative zeros.

Logarithmic Functions
For $x > 0, b > 0, b \neq 1$, the function $f(x) = \log_b x$ is known as the **logarithmic function** with base b. With $y = \log_b x$, its exponential equivalent is $b^y = x$. In either case, the exponent is y and the base is b. Therefore, $3 = \log_2 8$ is the same as $2^2 = 8$. So, in order to find the logarithm with base 2 of 8, find the exponent that when 2 is raised to that value results in 8. Similarly:

$$\log_3 243 = 5$$

In order to do this mentally, ask the question, what exponent does 3 need to be raised to that results in 243? The answer is 5. Most logarithms do not have whole number results. In this case, a calculator can be used. A calculator typically has buttons with base 10 and base e, so the change of base formula can be used to calculate these logs. For instance:

$$\log_3 55 = \frac{\log 55}{\log 3} = 3.64$$

Similarly, the natural logarithm with base e could be used to obtain the same result.

$$\log_3 55 = \frac{\ln 55}{\ln 3} = 3.64$$

The domain of a logarithmic function $f(x) = \log_b x$ is all positive real numbers. This is because the exponent must be a positive number. The range of a logarithmic function $f(x) = \log_b x$ is all real numbers. The graphs of all logarithmic functions of the form $f(x) = \log_b x$ always pass through the point (1, 0) because anything raised to the power of 0 is 1. Therefore, such a function always has an x-intercept at 1. If the base is greater than 1, the graph increases from the left to the right along the x-axis. If the base is between 0 and 1, the graph decreases from the left to the right along the x-axis. In both situations, the y-axis is a vertical **asymptote**. The graph will never touch the y-axis, but it does approach it closely.

Here are the graphs of the two cases of logarithmic functions:

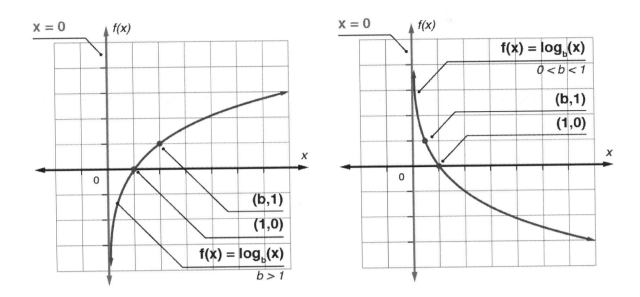

Finding and Applying Important Features of Graphs

A **graph** is a pictorial representation of the relationship between two variables. To read and interpret a graph, it is necessary to identify important features of the graph. First, read the title to determine what data sets are being related in the graph. Next, read the axis labels and understand the scale that is used. The horizontal axis often displays categories, like years, month, or types of pets. The vertical axis often displays numerical data like amount of income, number of items sold, or number of pets owned. Check to see what increments are used on each axis. The changes on the axis may represent fives, tens, hundreds, or any increment. Be sure to note what the increment is because it will affect the interpretation of the graph. Now, locate on the graph an element of interest and move across to find the element to which it relates. For example, notice an element displayed on the horizontal axis, find that element on the graph, and then follow it across to the corresponding point on the vertical axis. Using the appropriate scale, interpret the relationship.

Choosing Appropriate Graphs to Display Data

Data may be displayed with a line graph, bar graph, or pie chart.

- A line graph is used to display data that changes continuously over time.

- A bar graph is used to compare data from different categories or groups and is helpful for recognizing relationships.

- A pie chart is used when the data represents parts of a whole.

Data is often displayed with a line graph, bar graph, or pie chart.

The line graph below shows the number of push-ups that a student did over one week.

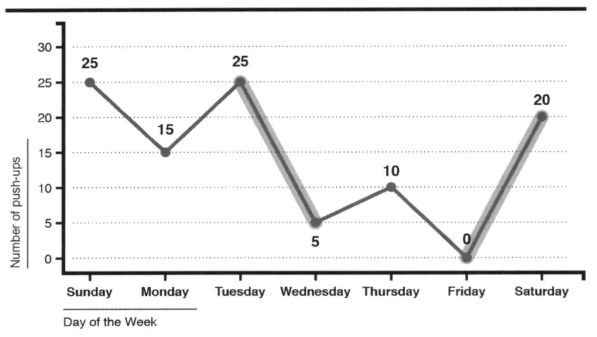

Notice that the horizontal axis displays the day of the week and the vertical axis displays the number of push-ups. A point is placed above each day of the week to show how many push-ups were done each day. For example, on Sunday the student did 25 push-ups. The line that connects the points shows how much the number of push-ups fluctuated throughout the week.

The bar graph below compares number of people who own various types of pets.

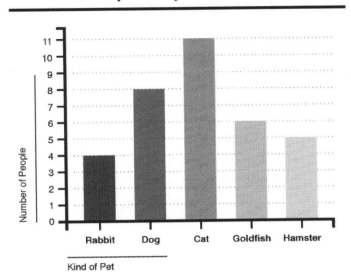

On the horizontal axis, the kind of pet is displayed. On the vertical axis, the number of people is displayed. Bars are drawn to show the number of people who own each type of pet. With the bar graph, it can quickly be determined that the fewest number of people own a rabbit and the greatest number of people own a cat.

The pie graph below displays students in a class who scored A, B, C, or D. Each slice of the pie is drawn to show the portion of the whole class that is represented by each letter grade. For example, the smallest portion represents students who scored a D. This means that the fewest number of students scored a D.

Student Grades

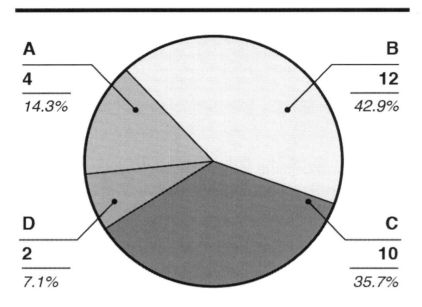

A
4
14.3%

B
12
42.9%

D
2
7.1%

C
10
35.7%

Geometry

Defining and Applying Knowledge of Shapes and Solids
Shapes are defined by their angles and number of sides. A shape with one continuous side, where all points on that side are equidistant from a center point is called a **circle.** A shape made with three straight line segments is a **triangle.** A shape with four sides is called a **quadrilateral,** but more specifically a **square, rectangle, parallelogram,** or **trapezoid,** depending on the interior angles. These shapes are two-dimensional and only made of straight lines and angles. **Solids** can be formed by combining these shapes and forming three-dimensional figures. These figures have another dimension because they add one more direction. Examples of solids include prisms or spheres. There are four figures below that can be described based on their sides and dimensions. Figure 1 is a **cone** because it has three dimensions, where the bottom is a circle and the top is formed by the sides combining to one point. Figure 2 is a **triangle** because it has two dimensions, made up of three-line segments.

Figure 3 is a **cylinder** made up of two base circles and a rectangle to connect them in three dimensions. Figure 4 is an **oval** because it is one continuous line in two dimensions, not equidistant from the center.

Shapes and Solids

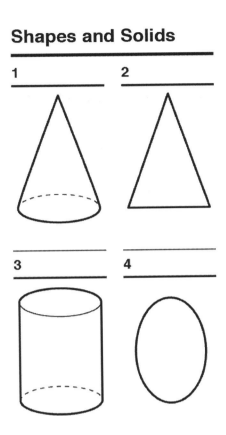

Figure 5 below is made up of squares in three dimensions, combined to make a **cube**. Figure 6 is a rectangle because it has four sides that intersect at right angles. More specifically, it can be described as a **square** because the four sides have equal measures. Figure 7 is a **pyramid** because the bottom shape is a square and the sides are all triangles. These triangles intersect at a point above the square. Figure 8 is a **circle** because it is made up of one continuous line where the points are all equidistant from one center point.

Shapes and Solids

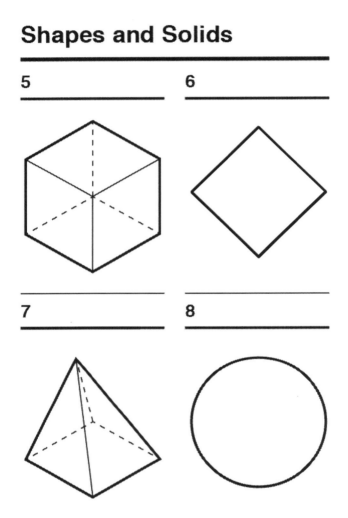

5

6

7

8

Congruence and Similarity

Two figures are **congruent** if they have the same shape and same size, meaning same angle measurements and equal side lengths. Two figures are **similar** if they have the same angle measurement but not side lengths. Basically, angles are congruent in similar triangles and their side lengths are constant multiplies of each other. Proving two shapes are similar involves showing that all angles are the same; proving two shapes are congruent involves showing that all angles are the same and that all sides are the same. If two pairs of angles are congruent in two triangles, then those triangles are similar because their third angle has to be equal due to the fact that all three angles add up to 180 degrees.

There are five main theorems that are used to show triangles are congruent. Each theorem involves showing different combinations of sides and angles are the same in two triangles, which proves the triangles are congruent. The **side-side-side (SSS) theorem** states that if all sides are equal in two triangles, the triangles are congruent. The **side-angle-side (SAS) theorem** states that if two pairs of sides are equal and the included angles are congruent in two triangles, then the triangles are congruent. Similarly, the **angle-side-angle (ASA) theorem** states that if two pairs of angles are congruent and the included side lengths are equal in two triangles, the triangles are similar. The **angle-angle-side (AAS) theorem** states that two triangles are congruent if they have two pairs of congruent angles and a pair of corresponding equal side lengths that are not included. Finally, the **hypotenuse-leg (HL) theorem** states that if two right triangles have equal hypotenuses and an equal pair of shorter sides, the triangles are congruent. An important item to note is that angle-angle-angle (AAA) is not enough information to have congruence because if three angles are equal in two triangles, the triangles can only be described as similar.

Using the Relationship Between Similarity, Right Triangles, and Trigonometric Ratios
Within two similar triangles, corresponding side lengths are proportional, and angles are equal. In other words, regarding corresponding sides in two similar triangles, the ratio of side lengths is the same. Recall that the SAS theorem for similarity states that if an angle in one triangle is congruent to an angle in a second triangle, and the lengths of the sides in both triangles are proportional, then the triangles are similar. Also, because the ratio of two sides in two similar right triangles is the same, the trigonometric ratios in similar right triangles are always going to be equal.

If two triangles are similar, and one is a right triangle, the other is a right triangle. The definition of similarity ensures that each triangle has a 90-degree angle. In a similar sense, if two triangles are right triangles containing a pair of equal acute angles, the triangles are similar because the third pair of angles must be equal as well. However, right triangles are not necessarily always similar.

The following triangles are similar:

Similar Triangles

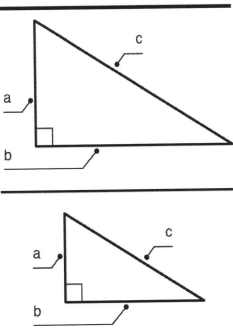

It is not always apparent at first glance, but theorems can be used to show similarity. The **Pythagorean Theorem** can be used to find the missing side lengths in both triangles. In the larger triangle, the missing side is the hypotenuse, *c*. Therefore:

$$9^2 + 12^2 = c^2$$

This equation is equivalent to $225 = c^2$, so taking the square root of both sides results in the positive root $c = 15$. In the other triangle, the Pythagorean Theorem can be used to find the missing side length *b*. The theorem shows that:

$$6^2 + b^2 = 10^2$$

and *b* is then solved for to obtain $b = 8$. The ratio of the sides in the larger triangle to the sides in the smaller triangle is the same value, 1.5. Therefore, the sides are proportional. Because they are both right triangles, they have a congruent angle. The SAS theorem for similarity can be used to show that these two triangles are similar.

Surface Area and Volume Measurements

Surface area is defined as the area of the surface of a figure. A **pyramid** has a surface made up of four triangles and one square. To calculate the surface area of a pyramid, the areas of each individual shape are calculated. Then the areas are added together. This method of decomposing the shape into two-dimensional figures to find area, then adding the areas, can be used to find surface area for any figure. Once these measurements are found, the area is described with square units. For example, the following figure shows a rectangular prism. The figure beside it shows the rectangular prism broken down into two-dimensional shapes, or rectangles. The area of each rectangle can be calculated by multiplying the length by the width. The area for the six rectangles can be represented by the following expression:

$$5 \times 6 + 5 \times 10 + 5 \times 6 + 6 \times 10 + 5 \times 10 + 6 \times 10$$

The total for all these areas added together is 280 m^2, or 280 square meters.

This measurement represents the surface area because it is the area of all six surfaces of the rectangular prism.

The Net of a Rectangular Prism

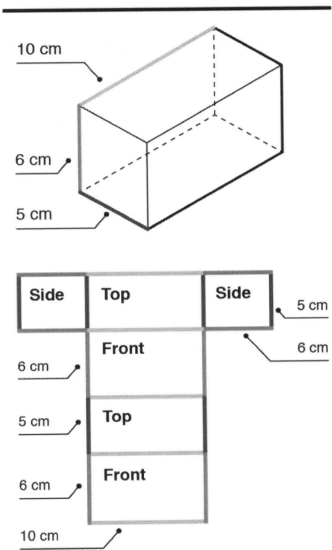

Surface area of three-dimensional figures is the total area of each of the faces of the figures. Nets are used to lay out each face of an object. The following figure shows a triangular prism. The bases are triangles and the sides are rectangles. The second figure shows the net for this triangular prism. The dimensions are labeled for each of the faces of the prism. The area for each of the two triangles can be determined by the formula:

$$A = \frac{1}{2}bh = \frac{1}{2} \times 8 \times 9 = 36 \text{ cm}^2$$

The rectangle areas can be described by the equation:

$$A = lw = 8 \times 5 + 9 \times 5 + 10 \times 5 = 40 + 45 + 50 = 135 \text{ cm}^2$$

The area for the triangles can be multiplied by two, then added to the rectangle areas to yield a total surface area of 207 cm².

A Triangular Prism and Its Net

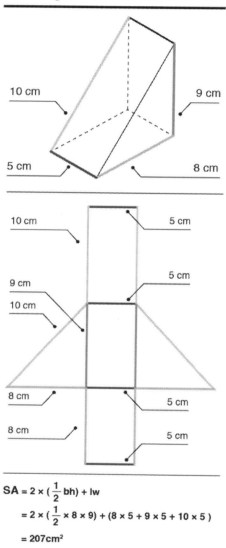

10 cm 9 cm

5 cm 8 cm

10 cm 5 cm

5 cm

9 cm

10 cm

8 cm 5 cm

8 cm 5 cm

$\text{SA} = 2 \times (\frac{1}{2} bh) + lw$

$= 2 \times (\frac{1}{2} \times 8 \times 9) + (8 \times 5 + 9 \times 5 + 10 \times 5)$

$= 207 \text{cm}^2$

Another shape that has a surface area is a cylinder. The shapes used to make up the **cylinder** are two circles and a rectangle wrapped around between the two circles. A common example of a cylinder is a can. The two circles that make up the bases are obvious shapes. The rectangle can be more difficult to see, but the label on a can will help illustrate it. When the label is removed from a can and laid flat, the shape is a rectangle. When the areas for each shape are needed, there will be two formulas. The first is the area for the circles on the bases. This area is given by the formula

$$A = \pi r^2$$

There will be two of these areas. Then the area of the rectangle must be determined. The width of the rectangle is equal to the height of the can, h. The length of the rectangle is equal to the circumference of the base circle, $2\pi r$. The area for the rectangle can be found by using the formula

$$A = 2\pi r \times h$$

By adding the two areas for the bases and the area of the rectangle, the surface area of the cylinder can be found, described in units squared.

Finding the Volume and Surface Area of Right Rectangular Prisms, Including Those with Fractional Edge Lengths

Right rectangular prisms are those prisms in which all sides are rectangles and all angles are right, or equal to 90 degrees. The volume for these objects can be found by multiplying the length by the width by the height. The formula is $V = lwh$. For the following prism, the volume formula is

$$V = 6\frac{1}{2} \times 3 \times 9$$

When dealing with fractional edge lengths, it is helpful to convert the length to an improper fraction. The length $6\frac{1}{2}$ cm becomes $\frac{13}{2}$ cm. Then the formula becomes:

$$V = \frac{13}{2} \times 3 \times 9$$

$$\frac{13}{2} \times \frac{3}{1} \times \frac{9}{1} = \frac{351}{2}$$

This value for volume is better understood when turned into a mixed number, which would be $175\frac{1}{2}$ cm^3.

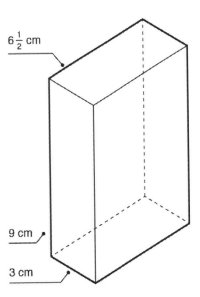

When dimensions for length are given with fractional parts, it can be helpful to turn the mixed number into an improper fraction, then multiply to find the volume, then convert back to a mixed number. When finding surface area, this conversion to improper fractions can also be helpful. The surface area can be found for the same prism above by breaking down the figure into basic shapes. These shapes are rectangles, made up of the two bases, two sides, and the front and back.

The formula for the surface area uses the area for each of these shapes for the terms in the following equation:

$$SA = 6\frac{1}{2} \times 3 + 6\frac{1}{2} \times 3 + 3 \times 9 + 3 \times 9 + 6\frac{1}{2} * 9 + 6\frac{1}{2} \times 9$$

Because there are so many terms in a surface area formula and because this formula contains a fraction, it can be simplified by combining groups that are the same. Each set of numbers is used twice, to represent areas for the opposite sides of the prism. The formula can be simplified to:

$$SA = 2\left(6\frac{1}{2} \times 3\right) + 2(3 \times 9) + 2\left(6\frac{1}{2} \times 9\right)$$

$$2\left(\frac{13}{2} \times 3\right) + 2(27) + 2\left(\frac{13}{2} \times 9\right)$$

$$2\left(\frac{39}{2}\right) + 54 + 2\left(\frac{117}{2}\right)$$

$$39 + 54 + 117 = 210 \text{ cm}^2$$

Understanding Composition of Objects

Composition of objects is the way objects are used in conjunction with each other to form bigger, more complex shapes. For example, a rectangle and a triangle can be used together to form an arrow. Arrows can be found in many everyday scenarios, but are often not seen as the composition of two different shapes. A square is a common shape, but it can also be the composition of shapes. As seen in the second figure, there are many shapes used in the making of the one square. There are five triangles that are three different sizes. There is also one square and one parallelogram used to compose this square. These shapes can be used to compose each more complex shape because they line up, side by side, to fill in the shape with no gaps. This defines composition of shapes where smaller shapes are used to make larger, more complex ones.

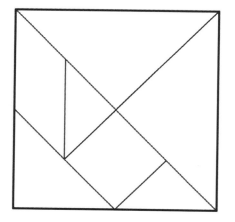
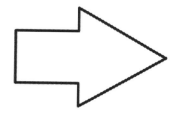

Solving for Missing Values in Triangles, Circles, and Other Figures

Solving for missing values in shapes requires knowledge of the shape and its characteristics. For example, a triangle has three sides and three angles that add up to 180 degrees. If two angle measurements are given, the third can be calculated. For the triangle below, the one given angle has a measure of 55 degrees. The missing angle is x. The third angle is labeled with a square, which indicates a measure of 90 degrees. Because all angles must sum to 180 degrees, the following equation can be used to find the missing x-value:

$$55° + 90° + x = 180°$$

Adding the two given angles and subtracting the total from 180, the missing angle is found to be 35 degrees.

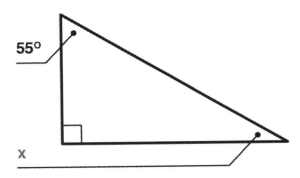

A similar problem can be solved with circles. If the radius is given but the circumference is unknown, the circumference can be calculated based on the formula $C = 2\pi r$. This example can be used in the figure below. The radius can be substituted for r in the formula. Then the circumference can be found as:

$$C = 2\pi \times 8 = 16\pi = 50.24 \text{ cm}$$

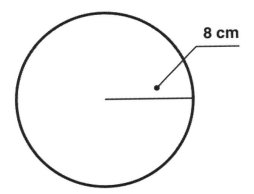

Other figures that may have missing values could be the length of a square, given the area, or the perimeter of a rectangle, given the length and width. All of the missing values can be found by first identifying all the characteristics that are known about the shape, then looking for ways to connect the missing value to the given information.

Using Trigonometric Ratios

Within right triangles, trigonometric ratios can be defined for the acute angle within the triangle. Consider the following right triangle. The side across from the right angle is known as the **hypotenuse,** the acute angle being discussed is labeled **θ,** the side across from the acute angle is known as the **opposite** side, and the other side is known as the **adjacent** side.

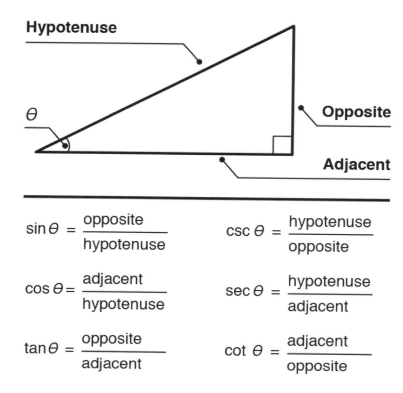

$$\sin \theta = \frac{\text{opposite}}{\text{hypotenuse}} \qquad \csc \theta = \frac{\text{hypotenuse}}{\text{opposite}}$$

$$\cos \theta = \frac{\text{adjacent}}{\text{hypotenuse}} \qquad \sec \theta = \frac{\text{hypotenuse}}{\text{adjacent}}$$

$$\tan \theta = \frac{\text{opposite}}{\text{adjacent}} \qquad \cot \theta = \frac{\text{adjacent}}{\text{opposite}}$$

The six trigonometric ratios are shown above as well. "Sin" is short for sine, "cos" is short for cosine, "tan" is short for tangent, "csc" is short for cosecant, "sec" is short for secant, and "cot" is short for cotangent. A mnemonic device exists that is helpful to remember the ratios. SOHCAHTOA stands for Sine = Opposite/Hypotenuse, Cosine = Adjacent/Hypotenuse, and Tangent = Opposite/Adjacent. The other three trigonometric ratios are reciprocals of sine, cosine, and tangent because:

$$\csc \theta = \frac{1}{\sin \theta}, \sec \theta = \frac{1}{\cos \theta}, \text{ and } \cot \theta = \frac{1}{\tan \theta}$$

The **Pythagorean Theorem** is an important relationship between the three sides of a right triangle. It states that the square of hypotenuse is equal to the sum of the squares of the other two sides. When using the Pythagorean Theorem, the hypotenuse is labeled as side c, the opposite is labeled as side a, and the adjacent side is side b.

The theorem can be seen in the following diagram:

The Pythagorean Theorem

$$a^2 + b^2 = c^2$$

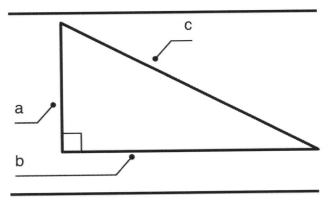

Both the trigonometric ratios and Pythagorean Theorem can be used in problems that involve finding either a missing side or missing angle of a right triangle. Look to see what sides and angles are given and select the correct relationship that will assist in finding the missing value. These relationships can also be used to solve application problems involving right triangles. Often, it is helpful to draw a figure to represent the problem to see what is missing.

Equations of Conic Sections
The intersection of a plane and a double right circular cone is called a **conic section**. There are four basic types of conic sections, a circle, a parabola, a hyperbola, and an ellipse. The equation of a **circle** is given by:

$$(x - h)^2 + (y - k)^2 = r^2$$

where the center of the circle is given by (h, k) and the radius of the circle is r. A parabola that opens up or down has a horizontal axis. The equation of a **parabola with a horizontal axis** is given by:

$$(y - k)^2 = 4p(x - h)$$

where $p \neq 0$ and the vertex is given by (h, k). A parabola that opens to the left or right has a vertical axis. The equation of the **parabola with a vertical axis** is given by:

$$(x - h)^2 = 4p(y - k)$$

where $p \neq 0$ and the vertex is given by (h, k). The equation of an **ellipse** with a horizontal major axis and center (h, k) is given by:

$$\frac{(x - h)^2}{a^2} + \frac{(y - k)^2}{b^2} = 1$$

The distance between center and either focus is c with:

$$c^2 = a^2 - b^2$$

when $a > b > 0$. The major axis has length $2a$ and the minor axis has length $2b$. For an ellipse with a vertical major axis and center (h, k), where $a > b > 0$, the a and b switch places so the equation is given by:

$$\frac{(x - h)^2}{b^2} + \frac{(y - k)^2}{a^2} = 1$$

The major axis still has length $2a$ and the minor axis still has length $2b$, and the distance between center and either focus is $c^2 = a^2 - b^2$, where $a > b > 0$.

A **hyperbola** has an equation similar to the ellipse except that there is a minus in place of the plus sign. A hyperbola with a vertical transverse axis has equation:

$$\frac{(x - h)^2}{a^2} - \frac{(y - k)^2}{b^2} = 1$$

A hyperbola with a horizontal transverse axis has equation:

$$\frac{(y - k)^2}{a^2} - \frac{(x - h)^2}{b^2} = 1$$

For each of these, the center is given by (h, k) and distance between the vertices $2a$.

Statistics & Probability

Describing Center and Spread of Distributions

One way information can be interpreted from tables, charts, and graphs is through statistics. The three most common calculations for a set of data are the mean, median, and mode. These three are called **measures of central tendency**. Measures of central tendency are helpful in comparing two or more different sets of data.

The **mean** refers to the average and is found by adding up all values and dividing the total by the number of values. In other words, the mean is equal to the sum of all values divided by the number of data entries. For example, if you bowled a total of 532 points in 4 bowling games, your mean score was:

$$\frac{532}{4} = 133 \text{ points per game}$$

A common application of mean useful to students is calculating what he or she needs to receive on a final exam to receive a desired grade in a class.

The **median** is found by lining up values from least to greatest and choosing the middle value. If there's an even number of values, then the mean of the two middle amounts must be calculated to find the median. For example, the median of the set of dollar amounts $5, $6, $9, $12, and $13 is $9. The median of the set of dollar amounts $1, $5, $6, $8, $9, $10 is $7, which is the mean of $6 and $8.

The **mode** is the value that occurs the most. The mode of the data set {1, 3, 1, 5, 5, 8, 10} actually refers to two numbers: 1 and 5. In this case, the data set is bimodal because it has two modes. A data set can have no mode if no amount is repeated. Another useful statistic is range.

The **range** for a set of data refers to the difference between the highest and lowest value.

In some cases, some numbers in a list of data might have weights attached to them. In that case, a **weighted mean** can be calculated. A common application of a weighted mean is GPA. In a semester, each class is assigned a number of credit hours, its weight, and at the end of the semester each student receives a grade. To compute GPA, an A is a 4, a B is a 3, a C is a 2, a D is a 1, and an F is a 0. Consider a student that takes a 4-hour English class, a 3-hour math class, and a 4-hour history class and receives all B's. The weighted mean, GPA, is found by multiplying each grade times its weight, number of credit hours, and dividing by the total number of credit hours.

Therefore, the student's GPA is:

$$\frac{3 \times 4 + 3 \times 3 + 3 \times 4}{11} = \frac{33}{1} = 3.0.$$

The following bar chart shows how many students attend a cycle class on each day of the week. To find the mean attendance for the week, each day's attendance can be added together:

$$10 + 7 + 6 + 9 + 8 + 14 + 4 = 58$$

and the total divided by the number of days:

$$58 \div 7 = 8.3$$

The mean attendance for the week was 8.3 people. The median attendance can be found by putting the attendance numbers in order from least to greatest: 4, 6, 7, 8, 9, 10, 14, and choosing the middle number: 8 people. The mode for attendance is none for this set of data because no numbers repeat. The range is 10, which is found by finding the difference between the lowest number, 4, and the highest number, 14.

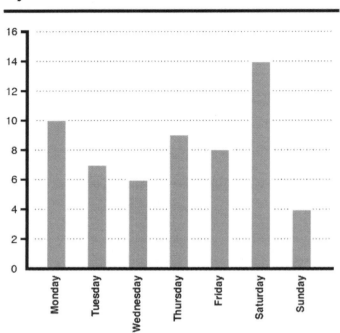

Cycle class attendance

A **histogram** is a bar graph used to group data into "bins" that cover a range on the horizontal, or x-axis. Histograms consist of rectangles whose height is equal to the frequency of a specific category. The horizontal axis represents the specific categories. Because they cover a range of data, these bins have no gaps between bars, unlike the bar graph above. In a histogram showing the heights of adult golden retrievers, the bottom axis would be groups of heights, and the y-axis would be the number of dogs in each range. Evaluating this histogram would show the height of most golden retrievers as falling within a certain range. It also provides information to find the average height and range for how tall golden retrievers may grow.

The following is a histogram that represents exam grades in a given class. The horizontal axis represents ranges of the number of points scored, and the vertical axis represents the number of students. For example, approximately 33 students scored in the 60 to 70 range.

Results of the exam

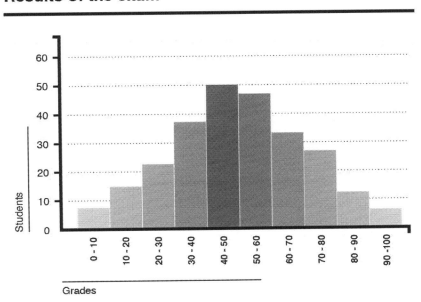

Certain measures of central tendency can be easily visualized with a histogram. If the points scored were shown with individual rectangles, the tallest rectangle would represent the mode. A bimodal set of data would have two peaks of equal height. Histograms can be classified as having data **skewed to the left, skewed to the right,** or **normally distributed**, which is also known as **bell-shaped**. These three classifications can be seen in the following chart:

Measures of central tendency images

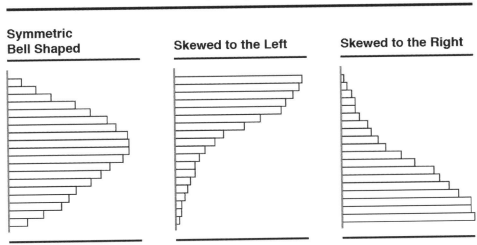

105

When the data is normal, the mean, median, and mode are all very close. They all represent the most typical value in the data set. The mean is typically used as the best measure of central tendency in this case because it does include all data points. However, if the data is skewed, the mean becomes less meaningful. The median is the best measure of central tendency because it is not affected by any outliers, unlike the mean. When the data is skewed, the mean is dragged in the direction of the skew. Therefore, if the data is not normal, it is best to use the median as the measure of central tendency.

The measures of central tendency and the range may also be found by evaluating information on a line graph.

The line graph shows the daily high and low temperatures:

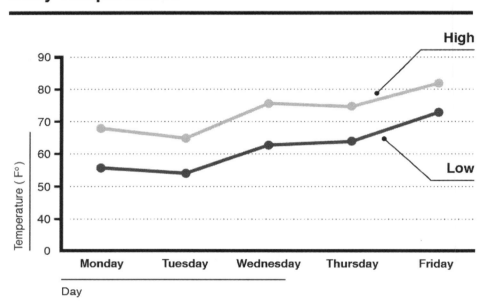

The average high temperature can be found by gathering data from each day on the triangle line. The days' highs are 82, 78, 75, 65, and 70. The average is found by adding them together to get 370, then dividing by 5 (because there are 5 temperatures). The average high for the five days is 74. If 74 degrees is found on the graph, then it falls in the middle of the values on the triangle line. The average low temperature can be found in the same way.

Given a set of data, the **correlation coefficient**, r, measures the association between all the data points. If two values are correlated, there is an association between them. However, correlation does not necessarily mean causation, or that one value causes the other. There is a common mistake made that assumes correlation implies causation. Average daily temperature and number of sunbathers are both correlated and have causation. If the temperature increases, that change in weather causes more people to want to catch some rays. However, wearing plus-size clothing and having heart disease are two variables that are correlated but do not have causation. The larger someone is, the more likely he or she is to have heart disease. However, being overweight does not cause someone to have the disease.

The value of the correlation coefficient is between −1 and 1, where −1 represents a perfect negative linear relationship, 0 represents no relationship between the two data sets, and 1 represents a perfect positive linear relationship. A negative linear relationship means that as *x*-values increase, *y*-values decrease. A positive linear relationship means that as *x* values increase, *y* values increase.

The formula for computing the correlation coefficient is:

$$r = \frac{n \sum xy - (\sum x)(\sum y)}{\sqrt{n(\sum x^2) - (\sum x)^2}\sqrt{n(\sum y^2) - (y)^2}}$$

where *n* is the number of data points. The closer *r* is to 1 or −1, the stronger the correlation. A correlation can be seen when plotting data. If the graph resembles a straight line, there is a correlation.

Applying and Analyzing Data Collection Methods

Data collection can be done through surveys, experiments, observations, and interviews. A **census** is a type of survey that is done with a whole population. Because it can be difficult to collect data for an entire population, sometimes, a sample survey is used. In this case, one would survey only a fraction of the population and make inferences about the data and generalizations about the larger population from which the sample was drawn. Sample surveys are not as accurate as a census, but this is an easier and less expensive method of collecting data. An **experiment** is used when a researcher wants to explain how one variable causes changes in another variable. For example, if a researcher wanted to know if a particular drug affects weight loss, he or she would choose a treatment group that would take the drug, and another group, the control group, that would not take the drug.

Special care must be taken when choosing these groups to ensure that bias is not a factor. **Bias** occurs when an outside factor influences the outcome of the research. In observational studies, the researcher does not try to influence either variable, but simply observes the behavior of the subjects. Interviews are sometimes used to collect data as well. The researcher will ask questions that focus on her area of interest in order to gain insight from the participants. When gathering data through observation or interviews, it is important that the researcher be well trained so that he or she does not influence the results and so that the study is reliable. A study is reliable if it can be repeated under the same conditions and the same results are received each time.

Understanding and Modeling Relationships in Bivariate Data

Independent and dependent are two types of variables that describe how they relate to each other. The **independent variable** is the variable controlled by the experimenter. It stands alone and isn't changed by other parts of the experiment. This variable is normally represented by *x* and is found on the horizontal, or *x*-axis, of a graph. The **dependent variable** changes in response to the independent variable. It reacts to, or depends on, the independent variable. This variable is normally represented by *y* and is found on the vertical, or *y*-axis of the graph.

The relationship between two variables, *x* and *y*, can be seen on a scatterplot.

The following scatterplot shows the relationship between weight and height. The graph shows the weight as *x* and the height as *y*. The first dot on the left represents a person who is 45 kg and approximately 150 cm tall. The other dots correspond in the same way. As the dots move to the right and weight increases, height also increases. A line could be drawn through the middle of the dots to move from bottom left to top right. This line would indicate a **positive correlation** between the variables. If the variables had a **negative correlation**, then the dots would move from the top left to the bottom right.

Height and Weight

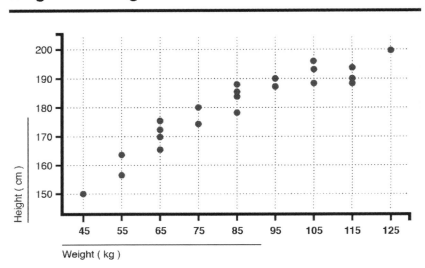

A **scatterplot** is useful in determining the relationship between two variables, but it's not required. Consider an example where a student scores a different grade on his math test for each week of the month. The independent variable would be the weeks of the month. The dependent variable would be the grades, because they change depending on the week. If the grades trended up as the weeks passed, then the relationship between grades and time would be positive. If the grades decreased as the time passed, then the relationship would be negative. (As the number of weeks went up, the grades went down.)

The relationship between two variables can further be described as strong or weak. The relationship between age and height shows a strong positive correlation because children grow taller as they get older. In adulthood, the relationship between age and height becomes weak, and the dots will spread out. People stop growing in adulthood, and their final heights vary depending on factors like genetics and health. The closer the dots on the graph, the stronger the relationship. As they spread apart, the relationship becomes weaker. If they are too spread out to determine a correlation up or down, then the variables are said to have no correlation.

Variables are values that change, so determining the relationship between them requires an evaluation of who or what changes them. If the variable changes because of a result in the experiment, then it's dependent. If the variable changes before the experiment, or is changed by the person controlling the experiment, then it's the independent variable. As they interact, one is manipulated by the other. The manipulator is the independent, and the manipulated is the dependent. Once the independent and dependent variable are determined, they can be evaluated to have a positive, negative, or no correlation.

Calculating Probabilities

Probability describes how likely it is that an event will occur. Probabilities are always a number from zero to 1. If an event has a high likelihood of occurrence, it will have a probability close to 1. If there is only a small chance that an event will occur, the likelihood is close to zero. A fair six-sided die has one of the numbers 1, 2, 3, 4, 5, and 6 on each side. When this die is rolled there is a one in six chance that it will land on 2. This is because there are six possibilities and only one side has a 2 on it. The probability then is $\frac{1}{6}$ or 0.167. The probability of rolling an even number from this die is three in six, which is $\frac{1}{2}$ or 0.5. This is because there are three sides on the die with even numbers (2, 4, 6), and there are six possible sides. The probability of rolling a number less than 10 is one because every side of the die has a number less than 6, so this is certain to occur. On the other hand, the probability of rolling a number larger than 20 is zero. There are no numbers greater than 20 on the die, so it is certain that this will not occur, thus the probability is zero.

If a teacher says that the probability of anyone passing her final exam is 0.2, is it highly likely that anyone will pass? No, the probability of anyone passing her exam is low because 0.2 is closer to zero than to 1. If another teacher is proud that the probability of students passing his class is 0.95, how likely is it that a student will pass? It is highly likely that a student will pass because the probability, 0.95, is very close to 1.

Using Two-Way Tables to Summarize Categorical Data and Relative Frequencies, and to Calculate Conditional Probability

A **two-way frequency table** displays categorical data with two variables, and it highlights relationships that exist between those two variables. Such tables are used frequently to summarize survey results, and are also known as **contingency tables**. Each cell shows a count pertaining to that individual variable paring, known as a **joint frequency**, and the totals of each row and column also are in the table. Consider the following two-way frequency table:

Distribution of the Residents of a Particular Village

	70 or older	69 or younger	Totals
Women	20	40	60
Men	5	35	40
Total	25	75	100

The table above shows the breakdown of ages and sexes of 100 people in a particular village. The total number of people in the data is shown in the bottom right corner. Each total is shown at the end of each row or column, as well. For instance, there were 25 people age 70 or older and 60 women in the data. The 20 in the first cell shows that out of 100 total villagers, 20 were women aged 70 or older. The 5 in the cell below shows that out of 100 total villagers, 5 were men aged 70 or older.

A two-way table can also show relative frequencies. If instead of the count, the percentage of people in each category was placed into the cells, the two-way table would show relative frequencies. If each frequency is calculated over the entire total of 100, the first cell would be 20% or 0.2. However, the relative frequencies can also be calculated over row or column totals. If row totals were used, the first cell would be:

$$\frac{20}{60} = 0.333 = 33.3\%$$

If column totals were used, the first cell would be:

$$\frac{20}{25} = 0.8 = 80\%$$

Such tables can be used to calculate **conditional probabilities**, which are probabilities that an event occurs, given another event. Consider a randomly selected villager. The probability of selecting a male 70 years old or older is $\frac{5}{100} = 0.05$ because there are 5 males over the age of 70 and 100 total villagers.

Integrating Essential Skills

Rates and Percentages

Ratios and Rates of Change
Recall that a **ratio** is the comparison of two different quantities. Comparing 2 apples to 3 oranges results in the ratio 2:3, which can be expressed as the fraction $\frac{2}{5}$. Note that order is important when discussing ratios. The number mentioned first is the antecedent, and the number mentioned second is the consequent. Note that the consequent of the ratio and the denominator of the fraction are *not* the same. When there are 2 apples to 3 oranges, there are five fruit total; two fifths of the fruit are apples, while three fifths are oranges. The ratio 2:3 represents a different relationship that the ratio 3:2. Also, it is important to make sure that when discussing ratios that have units attached to them, the two quantities use the same units. For example, to think of 8 feet to 4 yards, it would make sense to convert 4 yards to feet by multiplying by 3. Therefore, the ratio would be 8 feet to 12 feet, which can be expressed as the fraction $\frac{8}{20}$. Also, note that it is proper to refer to ratios in lowest terms. Therefore, the ratio of 8 feet to 4 yards is equivalent to the fraction $\frac{2}{5}$.

Therefore, the ratio of 8 feet to 4 yards is equivalent to the fraction $\frac{2}{3}$. Many real-world problems involve ratios. Often, problems with ratios involve proportions, as when two ratios are set equal to find the missing amount. However, some problems involve deciphering single ratios. For example, consider an amusement park that sold 345 tickets last Saturday. If 145 tickets were sold to adults and the rest of the tickets were sold to children, what would the ratio of the number of adult tickets to children's tickets be? A common mistake would be to say the ratio is 145:345. However, 345 is the total number of tickets sold.

There were 345 − 145 = 200 tickets sold to children. Thus, the correct ratio of adult to children's tickets is 145:200. As a fraction, this expression is written as $\frac{145}{200}$, which can be reduced to $\frac{29}{40}$.

While a ratio compares two measurements using the same units, **rates** compare two measurements with different units. Examples of rates would be $200 for 8 hours of work, or 500 miles traveled per 20 gallons. Because the units are different, it is important to always include the units when discussing rates. Rates can be easily seen because if they are expressed in words, the two quantities are usually split up using one of the following words: *for, per, on, from, in.* Just as with ratios, it is important to write rates in lowest terms. A common rate that can be found in many real-life situations is cost per unit. This quantity describes how much one item or one unit costs. This rate allows the best buy to be determined, given a couple of different sizes of an item with different costs. For example, if 2 quarts of soup was sold for $3.50 and 3 quarts was sold for $4.60, to determine the best buy, the cost per quart should be found.

$$\frac{\$3.50}{2 \text{ qt}} = \$1.75 \text{ per quart}$$

and

$$\frac{\$4.60}{3 \text{ qt}} = \$1.53 \text{ per quart}$$

Therefore, the better deal would be the 3-quart option.

Rate of change problems involve calculating a quantity per some unit of measurement. Usually the unit of measurement is time. For example, meters per second is a common rate of change. To calculate this measurement, find the distance traveled in meters and divide by total time traveled. The calculation is an average of the speed over the entire time interval.

Another common rate of change used in the real world is miles per hour. Consider the following problem that involves calculating an average rate of change in temperature. Last Saturday, the temperature at 1:00 a.m. was 34 degrees Fahrenheit, and at noon, the temperature had increased to 75 degrees Fahrenheit. What was the average rate of change over that time interval? The average rate of change is calculated by finding the total change in temperature and dividing by the total hours elapsed. Therefore, the rate of change was equal to:

$$\frac{75-34}{12-1} = \frac{41}{11} \text{ degrees per hour}$$

This quantity, rounded to two decimal places, is equal to 3.72 degrees per hour.

A common rate of change that appears in algebra is the slope calculation. Given a linear equation in one variable, $y = mx + b$, the **slope**, m, is equal to $\frac{rise}{run}$ or $\frac{change \ in \ y}{change \ in \ x}$. In other words, slope is equivalent to the ratio of the vertical and horizontal changes between any two points on a line. The vertical change is known as the **rise**, and the horizontal change is known as the **run**. Given any two points on a line (x_1, y_1) and (x_2, y_2), slope can be calculated with the formula:

$$m = \frac{y_2 - y_1}{x_2 - x_1} = \frac{\Delta y}{\Delta x}$$

Common real-world applications of slope include determining how steep a staircase should be, calculating how steep a road is, and determining how to build a wheelchair ramp.

Many times, problems involving rates and ratios involve proportions. A proportion states that two ratios (or rates) are equal. The property of cross products can be used to determine if a proportion is true, meaning both ratios are equivalent. If $\frac{a}{b} = \frac{c}{d}$, then to clear the fractions, multiply both sides by the least common denominator, bd. This results in $ad = bc$, which is equal to the result of multiplying along both diagonals. For example, $\frac{4}{40} = \frac{1}{10}$ grants the cross product:

$$4 \times 10 = 40 \times 1$$

which is equivalent to $40 = 40$ and shows that this proportion is true. Cross products are used when proportions are involved in real-world problems. Consider the following: If 3 pounds of fertilizer will cover 75 square feet of grass, how many pounds are needed for 375 square feet? To solve this problem, a proportion can be set up using two ratios. Let x equal the unknown quantity, pounds needed for 375 feet. Then, the equation found by setting the two given ratios equal to one another is:

$$\frac{3}{75} = \frac{x}{375}$$

Cross-multiplication gives:

$$3 \times 375 = 75x$$

Therefore, $1{,}125 = 75x$. Divide both sides by 75 to get $x = 15$. Therefore, 15 pounds of fertilizer are needed to cover 375 square feet of grass.

Another application of proportions involves similar triangles. If two triangles have the same measurement as two triangles in another triangle, the triangles are said to be **similar.** If two are the same, the third pair of angles are equal as well because the sum of all angles in a triangle is equal to 180 degrees. Each pair of equivalent angles are known as **corresponding angles. Corresponding sides** face the corresponding angles, and it is true that corresponding sides are in proportion.

For example, consider the following set of similar triangles:

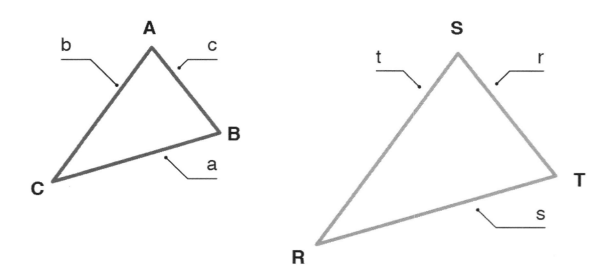

Angles A and S have the same measurement, angles C and R have the same measurement, and angles B and T have the same measurement. Therefore, the following proportion can be set up from the sides:

$$\frac{c}{r} = \frac{a}{s} = \frac{b}{t}$$

This proportion can be helpful in finding missing lengths in pairs of similar triangles. For example, if the following triangles are similar, a proportion can be used to find the missing side lengths, a and b.

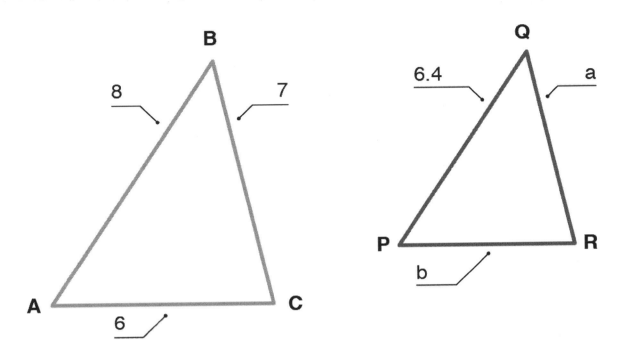

The proportions $\frac{8}{6.4} = \frac{6}{b}$ and $\frac{8}{6.4} = \frac{7}{a}$ can both be cross multiplied and solved to obtain $a = 5.6$ and $b = 4.8$.

A real-life situation that uses similar triangles involves measuring shadows to find heights of unknown objects. Consider the following problem: A building casts a shadow that is 120 feet long, and at the same time, another building that is 80 feet high casts a shadow that is 60 feet long. How tall is the first building? Each building, together with the sun rays and shadows casted on the ground, forms a triangle. They are similar because each building forms a right angle with the ground, and the sun rays form equivalent angles. Therefore, these two pairs of angles are both equal. Because all angles in a triangle add up to 180 degrees, the third angles are equal as well. Both shadows form corresponding sides of the triangle, the buildings form corresponding sides, and the sun rays form corresponding sides. Therefore, the triangles are similar, and the following proportion can be used to find the missing building length:

$$\frac{120}{x} = \frac{60}{80}$$

Cross-multiply to obtain the cross products, $9600 = 60x$. Then, divide both sides by 60 to obtain $x = 160$. This solution means that the other building is 160 feet high.

Percentages

Percentages are defined to be parts per one hundred. To convert a decimal to a percentage, move the decimal point two units to the right and place the percent sign after the number. Percentages appear in many scenarios in the real world. It is important to make sure the statement containing the percentage is translated to a correct mathematical expression. Be aware that it is extremely common to make a mistake when working with percentages within word problems.

An example of a word problem containing a percentage is the following: 35% of people speed when driving to work. In a group of 5,600 commuters, how many would be expected to speed on the way to their place of employment? The answer to this problem is found by finding 35% of 5,600. First, change the percentage to the decimal 0.35. Then compute the product:

$$0.35 \times 5,600 = 1,960$$

Therefore, it would be expected that 1,960 of those commuters would speed on their way to work based on the data given. In this situation, the word "of" signals to use multiplication to find the answer. Another way percentages are used is in the following problem: Teachers work 8 months out of the year. What percent of the year do they work? To answer this problem, find what percent of 12 the number 8 is, because there are 12 months in a year. Therefore, divide 8 by 12, and convert that number to a percentage:

$$\frac{8}{12} = \frac{2}{3} = 0.66\overline{6}$$

The percentage rounded to the nearest tenth place tells us that teachers work 66.7% of the year. Percentages also appear in real-world application problems involving finding missing quantities like in the following question: 60% of what number is 75? To find the missing quantity, an equation can be used. Let x be equal to the missing quantity.

Therefore:

$$0.60x = 75$$

Divide each side by 0.60 to obtain 125. Therefore, 60% of 125 is equal to 75.

Sales tax is an important application relating to percentages because tax rates are usually given as percentages. For example, a city might have an 8% sales tax rate. Therefore, when an item is purchased with that tax rate, the real cost to the customer is 1.08 times the price in the store. For example, a $25 pair of jeans costs the customer:

$$\$25 \times 1.08 = \$27$$

Sales tax rates can also be determined if they are unknown when an item is purchased. If a customer visits a store and purchases an item for $21.44, but the price in the store was $19, they can find the tax rate by first subtracting:

$$\$21.44 - \$19$$

to obtain $2.44, the sales tax amount. The sales tax is a percentage of the in-store price. Therefore, the tax rate is:

$$\frac{2.44}{19} = 0.128$$

which has been rounded to the nearest thousandths place. In this scenario, the actual sales tax rate given as a percentage is 12.8%.

Solving Unit Rate Problems

A **unit rate** is a rate with a denominator of one. It is a comparison of two values with different units where one value is equal to one. Examples of unit rates include 60 miles per hour and 200 words per minute. Problems involving unit rates may require some work to find the unit rate. For example, if Mary travels 360 miles in 5 hours, what is her speed, expressed as a unit rate? The rate can be expressed as the following fraction:

$$\frac{360 \; miles}{5 \; hours}$$

The denominator can be changed to one by dividing by five. The numerator will also need to be divided by five to follow the rules of equality. This division turns the fraction into:

$$\frac{72 \; miles}{1 \; hour}$$

which can now be labeled as a unit rate because one unit has a value of one. Another type question involves the use of unit rates to solve problems. For example, if Trey needs to read 300 pages and his average speed is 75 pages per hour, will he be able to finish the reading in 5 hours? The unit rate is 75 pages per hour, so the total of 300 pages can be divided by 75 to find the time. After the division, the time it takes to read is four hours. The answer to the question is yes, Trey will finish the reading within 5 hours.

Proportional Relationships

Fractions appear in everyday situations, and in many scenarios, they appear in the real-world as ratios and in proportions. A **ratio** is formed when two different quantities are compared. For example, in a group of 50 people, if there are 33 females and 17 males, the ratio of females to males is 33 to 17. This expression can be written in the fraction form as $\frac{33}{50}$, where the denominator is the sum of females and males, or by using the ratio symbol, 33:17. The order of the number matters when forming ratios. In the same setting, the ratio of males to females is 17 to 33, which is equivalent to $\frac{17}{50}$ or 17:33. A **proportion** is an equation involving two ratios. The equation $\frac{a}{b} = \frac{c}{d}$, or $a : b = c : d$ is a proportion, for real numbers a, b, c, and d. Usually, in one ratio, one of the quantities is unknown, and cross-multiplication is used to solve for the unknown. Consider:

$$\frac{1}{4} = \frac{x}{5}$$

To solve for x, cross-multiply to obtain:

$$5 = 4x$$

Divide each side by 4 to obtain the solution:

$$x = \frac{5}{4}$$

It is also true that percentages are ratios in which the second term is 100 minus the first term. For example, 65% is 65:35 or $\frac{65}{100}$. Therefore, when working with percentages, one is also working with ratios.

Real-world problems frequently involve proportions. For example, consider the following problem: If 2 out of 50 pizzas are usually delivered late from a local Italian restaurant, how many would be late out of 235 orders? The following proportion would be solved with x as the unknown quantity of late pizzas:

$$\frac{2}{50} = \frac{x}{235}$$

Cross multiplying results in:

$$470 = 50x$$

Divide both sides by 50 to obtain:

$$x = \frac{470}{50}$$

which in lowest terms is equal to $\frac{47}{5}$. In decimal form, this improper fraction is equal to 9.4. Because it does not make sense to answer this question with decimals (portions of pizzas do not get delivered) the answer must be rounded. Traditional rounding rules would say that 9 pizzas would be expected to be delivered late. However, to be safe, rounding up to 10 pizzas out of 235 would probably make more sense.

Area, Surface Area, and Volume

Perimeter and area are two commonly used geometric quantities that describe objects. **Perimeter** is the distance around an object. The perimeter of an object can be found by adding the lengths of all sides. Perimeter may be used in problems dealing with lengths around objects such as fences or borders. It may also be used in finding missing lengths, or working backwards. If the perimeter is given, but a length is missing, use subtraction to find the missing length. Given a square with side length s, the formula for perimeter is $P = 4s$. Given a rectangle with length l and width w, the formula for perimeter is:

$$P = 2l + 2w$$

The perimeter of a triangle is found by adding the three side lengths, and the perimeter of a trapezoid is found by adding the four side lengths. The units for perimeter are always the original units of length, such as meters, inches, miles, etc. When discussing a circle, the distance around the object is referred to as its **circumference,** not perimeter. The formula for circumference of a circle is $C = 2\pi r$, where r represents the radius of the circle. This formula can also be written as $C = d\pi$, where d represents the diameter of the circle.

Area is the two-dimensional space covered by an object. These problems may include the area of a rectangle, a yard, or a wall to be painted. Finding the area may be a simple formula, or it may require multiple formulas to be used together. The units for area are square units, such as square meters, square inches, and square miles. Given a square with side length s, the formula for its area is $A = s^2$.

Some other common shapes are shown below:

Shape	Formula	Graphic
Rectangle	$Area = length \times width$	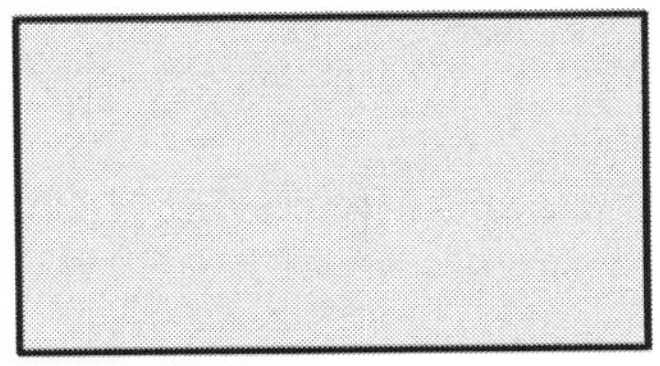
Triangle	$Area = \dfrac{1}{2} \times base \times height$	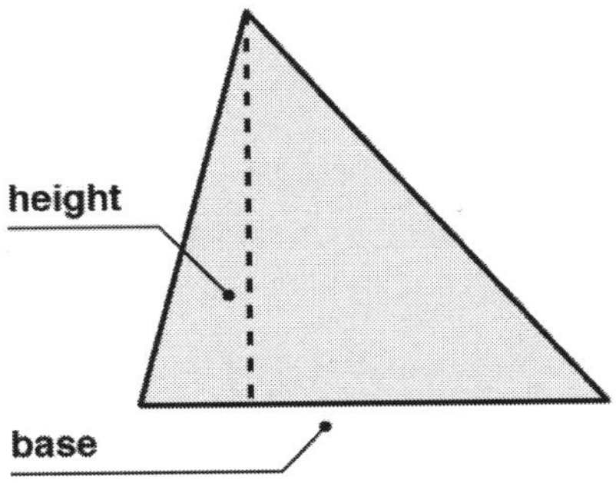 height ⟶ base
Circle	$Area = \pi \times radius^2$	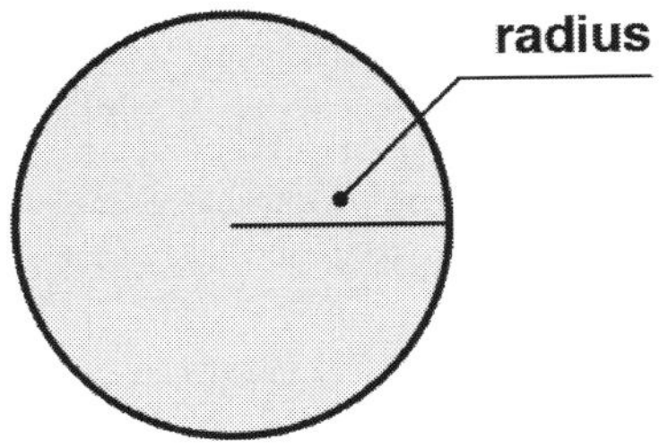 radius

The following formula, not as widely used as those shown above, but very important, is the area of a trapezoid:

Area of a Trapezoid

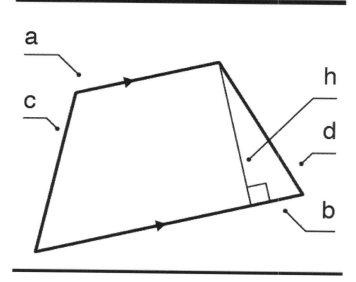

$$A = \frac{1}{2}\,(a + b)\,h$$

To find the area of the shapes above, use the given dimensions of the shape in the formula. Complex shapes might require more than one formula. To find the area of the figure below, break the figure into two shapes. The rectangle has dimensions 6 cm by 7 cm. The triangle has dimensions 6 cm by 6 cm. Plug the dimensions into the rectangle formula:

$$A = 6 \times 7$$

Multiplication yields an area of 42 cm². The triangle area can be found using the formula:

$$A = \frac{1}{2} \times 4 \times 6$$

Multiplication yields an area of 12 cm².

Add the areas of the two shapes to find the total area of the figure, which is 54 cm².

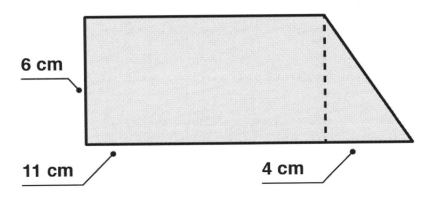

Instead of combining areas, some problems may require subtracting them, or finding the difference.

To find the area of the shaded region in the figure below, determine the area of the whole figure. Then subtract the area of the circle from the whole.

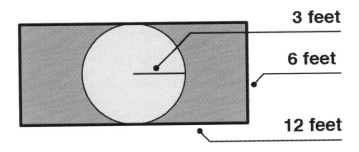

The following formula shows the area of the outside rectangle:

$$A = 12 \times 6 = 72 \text{ ft}^2$$

The area of the inside circle can be found by the following formula:

$$A = \pi(3)^2 = 9\pi = 28.3 \text{ ft}^2$$

As the shaded area is outside the circle, the area for the circle can be subtracted from the area of the rectangle to yield an area of 43.7 ft².

While some geometric figures may be given as pictures, others may be described in words. If a rectangular playing field with dimensions 95 meters long by 50 meters wide is measured for perimeter, the distance around the field must be found. The perimeter includes two lengths and two widths to measure the entire outside of the field. This quantity can be calculated using the following equation:

$$P = 2(95) + 2(50) = 290 \ m.$$

The distance around the field is 290 meters.

Perimeter and area are two-dimensional descriptions; volume is three-dimensional. **Volume** describes the amount of space that an object occupies, but it's different from area because it has three dimensions instead of two. The units for volume are cubic units, such as cubic meters, cubic inches, and cubic miles. Volume can be found by using formulas for common objects such as cylinders and boxes.

The following chart shows a diagram and formula for the volume of two objects.

Shape	Formula	Diagram
Rectangular Prism (box)	$V = length \times width \times height$	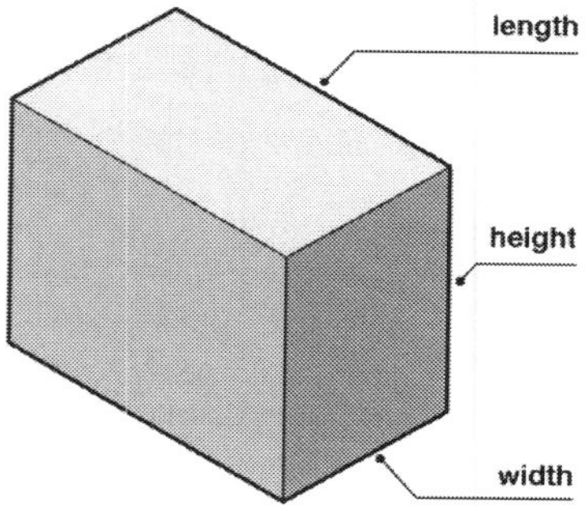
Cylinder	$V = \pi \times radius^2 \times height$	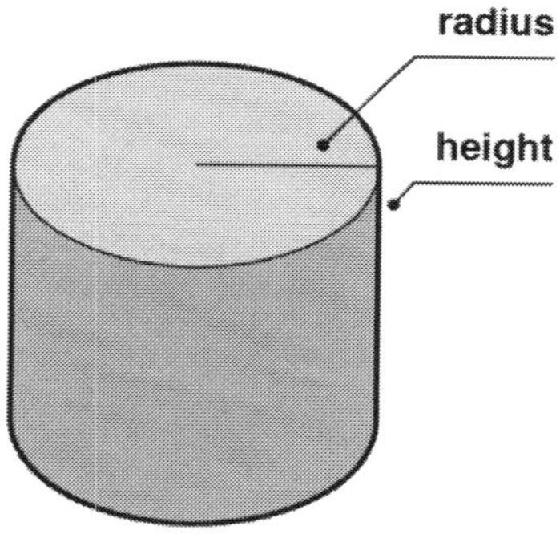

Volume formulas of these two objects are derived by finding the area of the bottom two-dimensional shape, such as the circle or rectangle, and then multiplying times the height of the three-dimensional shape. Other volume formulas include the volume of a cube with side length s:

$$V = s^3$$

the volume of a sphere with radius r:

$$V = \frac{4}{3}\pi r^3$$

120

and the volume of a cone with radius r and height h:

$$V = \frac{1}{3}\pi r^2 h$$

If a soda can has a height of 5 inches and a radius on the top of 1.5 inches, the volume can be found using one of the given formulas. A soda can is a cylinder. Knowing the given dimensions, the formula can be completed as follows:

$$V = \pi(radius)^2 \times height$$

$$\pi(1.5 \text{ in})^2 \times 5 \text{ in} = 35.325 \text{ in}^3$$

Notice that the units for volume are inches cubed because it refers to the number of cubic inches required to fill the can.

With any geometric calculations, it's important to determine what dimensions are given and what quantities the problem is asking for. If a connection can be made between them, the answer can be found.

Other geometric quantities can include angles inside a triangle. The sum of the measures of three angles in any triangle is 180 degrees. Therefore, if only two angles are known inside a triangle, the third can be found by subtracting the sum of the two known quantities from 180. Two angles whose sum is equal to 90 degrees are known as **complementary angles.** For example, angles measuring 72 and 18 degrees are complementary, and each angle is a complement of the other. Finally, two angles whose sum is equal to 180 degrees are known as **supplementary angles.** To find the supplement of an angle, subtract the given angle from 180 degrees. For example, the supplement of an angle that is 50 degrees is 180 − 50 = 130 degrees.

These terms involving angles can be seen in many types of word problems. For example, consider the following problem: The measure of an angle is 60 degrees less than two times the measure of its complement. What is the angle's measure? To solve this, let x be the unknown angle. Therefore, its complement is $90 - x$. The problem gives that:

$$x = 2(90 - x) - 60$$

To solve for x, distribute the 2, and collect like terms. This process results in:

$$x = 120 - 2x$$

Then, use the addition property to add $2x$ to both sides to obtain:

$$3x = 120$$

Finally, use the multiplication properties of equality to divide both sides by 3 to get $x = 40$. Therefore, the angle measures 40 degrees. Also, its complement measures 50 degrees.

Determining Surface Area Measurements

As mentioned previously, **surface area** is defined as the area of the surface of a figure. A pyramid has a surface made up of four triangles and one square. To calculate the surface area of a pyramid, the areas of each individual shape are calculated. Then the areas are added together. This method of decomposing the shape into two-dimensional figures to find area, then adding the areas, can be used to find surface area for any figure. Once these measurements are found, the area is described with square units. The surface area of a cube is found by multiplying the area of one side by six since there are six equivalent sides. For example, if the side length is 4 inches, the surface area is found using:

$$SA = 6 \times (4 \times 4) = 96 \; square \; inches$$

Average and Median

A data set can be described by calculating the mean, median, and mode. These values, called **measures of center,** allow the data to be described with a single value that is representative of the data set.

Again, the most common measure of center is the **mean,** also referred to as the **average.**

To calculate the mean,

- Add all data values together

- Divide by the sample size (the number of data points in the set)

The **median** is middle data value, so that half of the data lies below this value and half lies below the data value.

To calculate the median,

- Order the data from least to greatest

- The point in the middle of the set is the median

    o In the event that there is an even number of data points, add the two middle points and divide by 2

The **mode** is the data value that occurs most often.

To calculate the mode,

- Order the data from least to greatest

- Find the value that occurs most often

Example: Amelia is a leading scorer on the school's basketball team. The following data set represents the number of points that Amelia has scored in each game this season. Use the mean, median, and mode to describe the data.

16, 12, 26, 14, 28, 14, 12, 15, 25

Solution:

Mean:

$$16 + 12 + 26 + 14 + 28 + 14 + 12 + 15 + 25 = 162$$

$$162 \div 9 = 18$$

Amelia averages 18 points per game.

Median:

12, 12, 14, 14, **15**, 16, 25, 26, 28

Amelia's median score is 15.

Mode:

12, 12, 14, 14, 15, 16, 25, 26, 28

The numbers 12 and 14 each occur twice in the data set, so this set has 2 modes: 12 and 14.

The **range** is the difference between the largest and smallest values in the set. In the example above, the range is:

$$28 - 12 = 16$$

Expressing Numbers in Different Ways

Converting Non-Negative Fractions, Decimals, and Percentages

Within the number system, different forms of numbers can be used. It is important to be able to recognize each type, as well as work with, and convert between, the given forms. The **real number system** comprises natural numbers, whole numbers, integers, rational numbers, and irrational numbers. Natural numbers, whole numbers, integers, and irrational numbers typically are not represented as fractions, decimals, or percentages. Rational numbers, however, can be represented as any of these three forms. A **rational number** is a number that can be written in the form $\frac{a}{b}$, where a and b are integers, and b is not equal to zero. In other words, rational numbers can be written in a fraction form. The value a is the **numerator**, and b is the **denominator**. If the numerator is equal to zero, the entire fraction is equal to zero. Non-negative fractions can be less than 1, equal to 1, or greater than 1. Fractions are less than 1 if the numerator is smaller (less than) than the denominator. For example, $\frac{3}{4}$ is less than 1. A fraction is equal to 1 if the numerator is equal to the denominator.

For instance, $\frac{4}{4}$ is equal to 1. Finally, a fraction is greater than 1 if the numerator is greater than the denominator: the fraction $\frac{11}{4}$ is greater than 1. When the numerator is greater than the denominator, the fraction is called an **improper fraction**.

An improper fraction can be converted to a **mixed number,** a combination of both a whole number and a fraction. To convert an improper fraction to a mixed number, divide the numerator by the denominator. Write down the whole number portion, and then write any remainder over the original denominator. For example, $\frac{11}{4}$ is equivalent to $2\frac{3}{4}$.

Conversely, a mixed number can be converted to an improper fraction by multiplying the denominator by the whole number and adding that result to the numerator.

Fractions can be converted to decimals. With a calculator, a fraction is converted to a decimal by dividing the numerator by the denominator. For example:

$$\frac{2}{5} = 2 \div 5 = 0.4$$

Sometimes, rounding might be necessary. Consider:

$$\frac{2}{7} = 2 \div 7 = 0.28571429$$

This decimal could be rounded for ease of use, and if it needed to be rounded to the nearest thousandth, the result would be 0.286. If a calculator is not available, a fraction can be converted to a decimal manually. First, find a number that, when multiplied by the denominator, has a value equal to 10, 100, 1,000, etc. Then, multiply both the numerator and denominator times that number. The decimal form of the fraction is equal to the new numerator with a decimal point placed as many place values to the left as there are zeros in the denominator. For example, to convert $\frac{3}{5}$ to a decimal, multiply both the numerator and denominator times 2, which results in $\frac{6}{10}$. The decimal is equal to 0.6 because there is one zero in the denominator, and so the decimal place in the numerator is moved one unit to the left.

In the case where rounding would be necessary while working without a calculator, an approximation must be found. A number close to 10, 100, 1,000, etc. can be used. For example, to convert $\frac{1}{3}$ to a decimal, the numerator and denominator can be multiplied by 33 to turn the denominator into approximately 100, which makes for an easier conversion to the equivalent decimal. This process results in $\frac{33}{99}$ and an approximate decimal of 0.33. Once in decimal form, the number can be converted to a percentage. Multiply the decimal by 100 and then place a percent sign after the number. For example, 0.614 is equal to 61.4%. In other words, move the decimal place two units to the right and add the percentage symbol.

Composing and Decomposing Multidigit Numbers

Composing and decomposing numbers reveals the place value held by each number 0 through 9 in each position. For example, the number 17 is read as "seventeen." It can be decomposed into the numbers 10 and 7. It can be described as 1 group of ten and 7 ones. The one in the tens place represents one set of ten. The seven in the ones place represents seven sets of one. Added together, they make a total of seventeen. The number 48 can be written in words as "forty-eight." It can be decomposed into the numbers 40 and 8, where there are 4 groups of ten and 8 groups of one. The number 296 can be decomposed into 2 groups of one hundred, 9 groups of ten, and 6 groups of one. There are two hundreds, nine tens, and six ones. Decomposing and composing numbers lays the foundation for visually picturing the number and its place value, and adding and subtracting multiple numbers with ease.

Modeling

Producing, Interpreting, Understanding, Evaluating, and Improving Models

Interpreting Relevant Information from Tables, Charts, and Graphs

Tables, charts, and graphs can be used to convey information about different variables. They are all used to organize, categorize, and compare data, and they all come in different shapes and sizes. Each type has its own way of showing information, whether it is in a column, shape, or picture. To answer a question relating to a table, chart, or graph, some steps should be followed. First, the problem should be read thoroughly to determine what is being asked to determine what quantity is unknown. Then, the title of the table, chart, or graph should be read. The title should clarify what data is actually being summarized in the table. Next, look at the key and labels for both the horizontal and vertical axes, if they are given. These items will provide information about how the data is organized. Finally, look to see if there is any more labeling inside the table. Taking the time to get a good idea of what the table is summarizing will be helpful as it is used to interpret information.

Tables are a good way of showing a lot of information in a small space. The information in a table is organized in columns and rows. For example, a table may be used to show the number of votes each candidate received in an election. By interpreting the table, one may observe which candidate won the election and which candidates came in second and third. In using a bar chart to display monthly rainfall amounts in different countries, rainfall can be compared between counties at different times of the year. Graphs are also a useful way to show change in variables over time, as in a line graph, or percentages of a whole, as in a pie graph.

The table below relates the number of items to the total cost. The table shows that 1 item costs $5. By looking at the table further, 5 items cost $25, 10 items cost $50, and 50 items cost $250. This cost can be extended for any number of items. Since 1 item costs $5, then 2 items would cost $10. Though this information isn't in the table, the given price can be used to calculate unknown information.

Number of Items	1	5	10	50
Cost ($)	5	25	50	250

A bar graph is a graph that summarizes data using bars of different heights. It is useful when comparing two or more items or when seeing how a quantity changes over time. It has both a horizontal and vertical axis. Interpreting **bar graphs** includes recognizing what each bar represents and connecting that to the two variables. The bar graph below shows the scores for six people on three different games. The color of the bar shows which game each person played, and the height of the bar indicates their score for that game. William scored 25 on game 3, and Abigail scored 38 on game 3. By comparing the bars, it's obvious that Williams scored lower than Abigail.

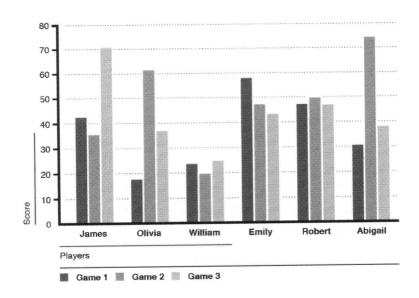

A line graph is a way to compare two variables. Each variable is plotted along an axis, and the graph contains both a horizontal and a vertical axis. On a **line graph,** the line indicates a continuous change. The change can be seen in how the line rises or falls, known as its slope, or rate of change. Often, in line graphs, the horizontal axis represents a variable of time. Audiences can quickly see if an amount has grown or decreased over time. The bottom of the graph, or the x-axis, shows the units for time, such as days, hours, months, etc. If there are multiple lines, a comparison can be made between what the two lines represent. For example, as shown previously, the following line graph shows the change in temperature over five days. The top line represents the high, and the bottom line represents the low for each day. Looking at the top line alone, the high decreases for a day, then increases on Wednesday. Then it decreased on Thursday and increases again on Friday. The low temperatures have a similar trend, shown in bottom line. The range in temperatures each day can also be calculated by finding the difference between the top line and bottom line on a particular day. On Wednesday, the range was 14 degrees, from 62 to 76° F.

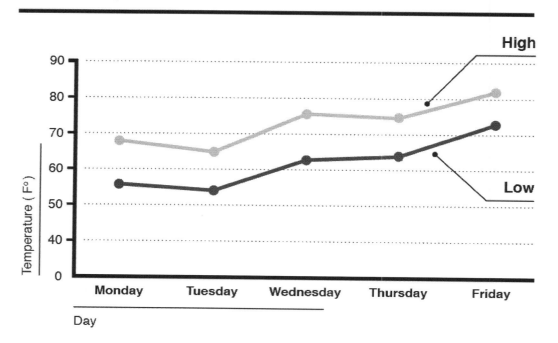

Daily Temperatures

Pie charts are used to show percentages of a whole, as each category is given a piece of the pie, and together all the pieces make up a whole. They are a circular representation of data which are used to highlight numerical proportion. It is true that the arc length of each pie slice is proportional to the amount it individually represents. When a pie chart is shown, an audience can quickly make comparisons by comparing the sizes of the pieces of the pie. They can be useful for comparison between different categories. The following pie chart is a simple example of three different categories shown in comparison to each other.

Light gray represents cats, dark gray represents dogs, and the gray between those two represents other pets. As the pie is cut into three equal pieces, each value represents just more than 33 percent, or $\frac{1}{3}$ of the whole. Values 1 and 2 may be combined to represent $\frac{2}{3}$ of the whole. In an example where the total pie represents 75,000 animals, then cats would be equal to $\frac{1}{3}$ of the total, or 25,000. Dogs would equal 25,000 and other pets would also equal 25,000.

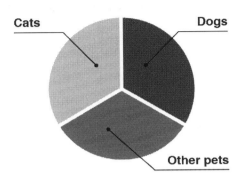

The fact that a circle is 360 degrees is used to create a pie chart. Because each piece of the pie is a percentage of a whole, that percentage is multiplied times 360 to get the number of degrees each piece represents. In the example above, each piece is $\frac{1}{3}$ of the whole, so each piece is equivalent to 120 degrees. Together, all three pieces add up to 360 degrees.

Stacked bar graphs, also used fairly frequently, are used when comparing multiple variables at one time. They combine some elements of both pie charts and bar graphs, using the organization of bar graphs and the proportionality aspect of pie charts. The following is an example of a stacked bar graph that represents the number of students in a band playing drums, flute, trombone, and clarinet. Each bar graph is broken up further into girls and boys.

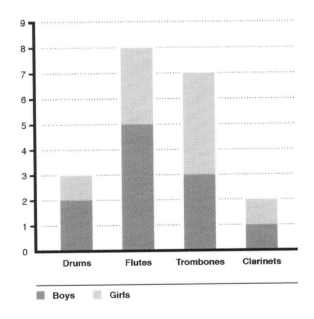

127

To determine how many boys play trombone, refer to the darker portion of the trombone bar, resulting in 3 students.

A **scatterplot** is another way to represent paired data. It uses Cartesian coordinates, like a line graph, meaning it has both a horizontal and vertical axis. Each data point is represented as a dot on the graph. The dots are never connected with a line. For example, the following is a scatterplot showing people's age versus height.

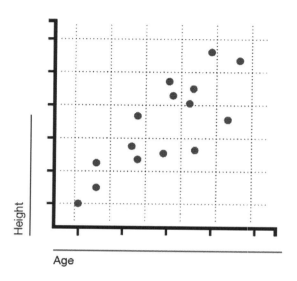

A scatterplot, also known as a **scattergram**, can be used to predict another value and to see if an association, known as a **correlation**, exists between a set of data. If the data resembles a straight line, the data is **associated.** The following is an example of a scatterplot in which the data does not seem to have an association:

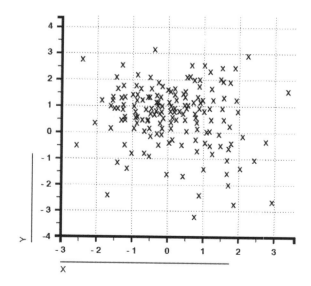

Sets of numbers and other similarly organized data can also be represented graphically. Venn diagrams are a common way to do so. A **Venn diagram** represents each set of data as a circle. The circles overlap, showing that each set of data is overlapping. A Venn diagram is also known as a **logic diagram** because it visualizes all possible logical combinations between two sets. Common elements of two sets are represented by the area of overlap. The following is an example of a Venn diagram of two sets A and B:

Parts of the Venn Diagram

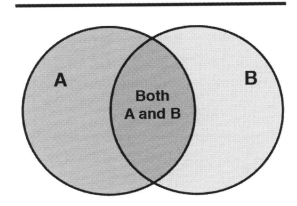

Another name for the area of overlap is the **intersection.** The intersection of A and B, $A \cap B$, contains all elements that are in both sets A and B. The **union** of A and B, $A \cup B$, contains all elements that are in either set A or set B. Finally, the **complement** of $A \cup B$ is equal to all elements that are not in either set A or set B. These elements are placed outside of the circles.

The following is an example of a Venn diagram in which 30 students were surveyed asking which type of siblings they had: brothers, sisters, or both. Ten students only had a brother, 7 students only had a sister, and 5 had both a brother and a sister. This number 5 is the intersection and is placed where the circles overlap. Two students did not have a brother or a sister. Two is therefore the complement and is placed outside of the circles.

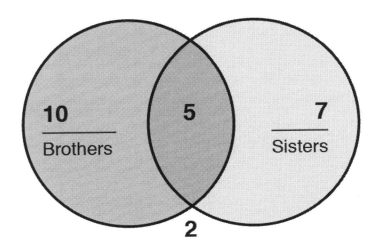

Venn diagrams can have more than two sets of data. The more circles, the more logical combinations are represented by the overlapping. The following is a Venn diagram that represents a different situation. Now, there were 30 students surveyed about the color of their socks. The innermost region represents those students that have green, pink, and blue socks on (perhaps a striped pattern). Therefore, 2 students had all three colors on their socks. In this example, all students had at least one of the three colors on their socks, so no one exists in the complement.

30 students

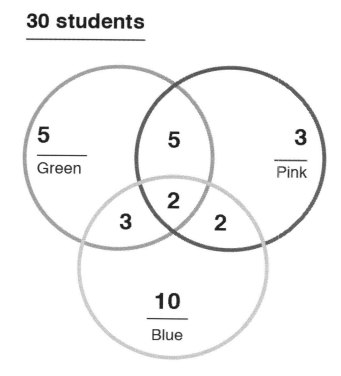

Venn diagrams are typically not drawn to scale, but if they are and their area is proportional to the amount of data it represents, it is known as an *area-proportional* Venn diagram.

Producing, Interpreting, Understanding, Evaluating, and Improving Models
Concrete models create a way of thinking about math that generates learning on a more permanent level. Memorizing abstract math formulas will not create a lasting effect on the brain. The following picture shows fractions represented by Lego blocks. By starting with the whole block of eight, it can be split into half, which is a four-block, and a fourth, which is a two-block. The one-eighth representation is a single block.

After splitting these up, addition and subtraction can be performed by adding or taking away parts of the blocks. Different combinations of fractions can be used to make a whole, or taken away to make various parts of a whole.

Using Colored Blocks to Model Functions

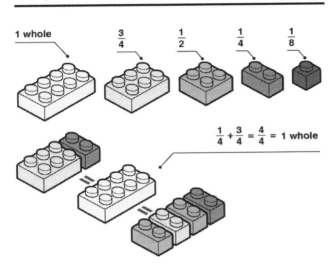

Multiplication can also be done using array models. The following picture shows a model of multiplying 3 times 4. **Arrays** are formed when the first factor is shown in a row. The second factor is shown in that number of columns. When the rectangle is formed, the blocks fill in to make a total, or the result of multiplication. The three rows and four columns show each factor and when the blocks are filled in, the total is twelve. Arrays are great ways to represent multiplication because they show each factor, and where the total comes from, with rows and columns.

Multiplication Array Model for 3 x 4

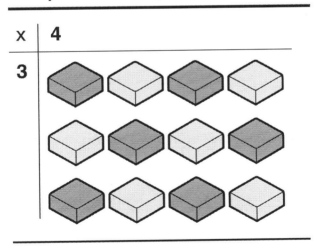

3 x 4 = 12

Another representation of fractions is shown below in the pie charts. Moving from whole numbers to part of numbers with fractions can be a concept that is difficult to grasp. Starting with a whole pie and splitting

131

it into parts can be helpful with generating fractions. The first pie shows quarters or sections that are one-fourth because it is split into four parts. The second pie shows parts that equal one-fifth because it is split into five parts. Pies can also be used to add fractions. If $\frac{1}{5}$ and $\frac{1}{4}$ are being added, a common denominator must be found by splitting the pies into the same number of parts. The same number of parts can be found by determining the least common multiple. For 4 and 5, the least common multiple is 20. The pies can be split until there are 20 parts. The same portion of the pie can be shaded for each fraction and then added together to find the sum.

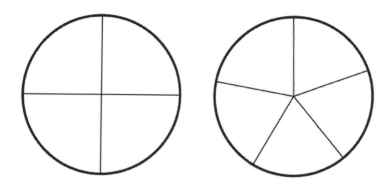

Illustrating and Explaining Multiplication and Division Problems Using Equations, Rectangular Arrays, and Area Models

Multiplication and division can be represented by equations. These equations show the numbers involved and the operation. For example, "eight multiplied by six equals forty-eight" is seen in the following equation: $8 \times 6 = 48$. This operation can be modeled by rectangular arrays where one factor, 8, is the number of rows, and the other factor, 6, is the number of columns, as follows:

Array of 8 x 6 = 48

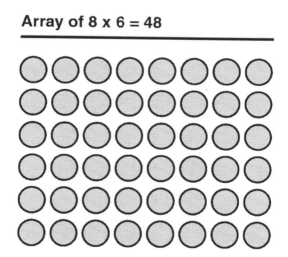

Rectangular arrays show what happens with the concept of multiplication. As one row of dots is drawn, that represents the first factor in the problem. Then the second factor is used to add the number of columns. The final model includes six rows of eight columns which results in forty-eight dots. These

132

rectangular arrays show how multiplication of whole numbers will result in a number larger than the factors.

Division can also be represented by equations and area models. A division problem such as "twenty-four divided by three equals eight" can be written as the following equation: 24 ÷ 8 = 3. The object below shows an area model to represent the equation. As seen in the model, the whole box represents 24 and the 3 sections represent the division by 3. In more detail, there could be 24 dots written in the whole box and each box could have 8 dots in it. Division shows how numbers can be divided into groups. For the example problem, it is asking how many numbers will be in each of the 3 groups that together make 24. The answer is 8 in each group.

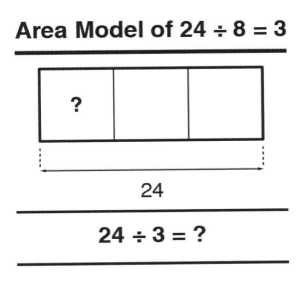

Area Model of 24 ÷ 8 = 3

24

24 ÷ 3 = ?

Converting Within and Between Standard and Metric Systems

When working with dimensions, sometimes the given units don't match the formula, and conversions must be made. The metric system has base units of meter for length, kilogram for mass, and liter for liquid volume. This system expands to three places above the base unit and three places below. These places correspond with prefixes with a base of 10.

The following table shows the conversions:

kilo-	hecto-	deca-	base	deci-	centi-	milli-
1,000 times the base	100 times the base	10 times the base		1/10 times the base	1/100 times the base	1/1000 times the base

To convert between units within the metric system, values with a base ten can be multiplied. The decimal can also be moved in the direction of the new unit by the same number of zeros on the number. For example, 3 meters is equivalent to 0.003 kilometers. The decimal moved three places (the same number of zeros for kilo-) to the left (the same direction from base to kilo-). Three meters is also equivalent to 3,000 millimeters. The decimal is moved three places to the right because the prefix milli- is three places to the right of the base unit.

The English Standard system used in the United States has a base unit of foot for length, pound for weight, and gallon for liquid volume. These conversions aren't as easy as the metric system because they aren't a base ten model. The following table shows the conversions within this system.

Length	Weight	Capacity
1 foot (ft) = 12 inches (in) 1 yard (yd) = 3 feet 1 mile (mi) = 5280 feet 1 mile = 1760 yards	1 pound (lb) = 16 ounces (oz) 1 ton = 2000 pounds	1 tablespoon (tbsp) = 3 teaspoons (tsp) 1 cup (c) = 16 tablespoons 1 cup = 8 fluid ounces (oz) 1 pint (pt) = 2 cups 1 quart (qt) = 2 pints 1 gallon (gal) = 4 quarts

When converting within the English Standard system, most calculations include a conversion to the base unit and then another to the desired unit. For example, take the following problem:

$$3 \text{ qt} = \underline{\quad} \text{ c}$$

There is no straight conversion from quarts to cups, so the first conversion is from quarts to pints. There are 2 pints in 1 quart, so there are 6 pints in 3 quarts. This conversion can be solved as a proportion:

$$\frac{3 \text{ qt}}{x} = \frac{1 \text{ qt}}{2 \text{ pt}}$$

It can also be observed as a ratio 2:1, expanded to 6:3. Then the 6 pints must be converted to cups. The ratio of pints to cups is 1:2, so the expanded ratio is 6:12. For 6 pints, the measurement is 12 cups. This problem can also be set up as one set of fractions to cancel out units. It begins with the given information and cancels out matching units on top and bottom to yield the answer. Consider the following expression:

$$\frac{3 \text{ qt}}{1} \times \frac{2 \text{ pt}}{1 \text{ qt}} \times \frac{2 \text{ c}}{1 \text{ pt}}$$

It's set up so that units on the top and bottom cancel each other out:

$$\frac{3 \text{ \cancel{qt}}}{1} \times \frac{2 \text{ \cancel{pt}}}{1 \text{ \cancel{qt}}} \times \frac{2 \text{ c}}{1 \text{ \cancel{pt}}}$$

The numbers can be calculated as $3 \times 2 \times 2$ on the top and 1 on the bottom. It still yields an answer of 12 cups.

This process of setting up fractions and canceling out matching units can be used to convert between standard and metric systems. A few common equivalent conversions are 2.54 cm = 1 in, 3.28 ft = 1 m, and 2.205 lb = 1 kg. Writing these as fractions allows them to be used in conversions. For the fill-in-the-blank problem 5 m = ___ ft, an expression using conversions starts with the expression:

$$\frac{5 \text{ m}}{1} \times \frac{3.28 \text{ ft}}{1 \text{ m}}$$

where the units of meters will cancel each other out and the final unit is feet. Calculating the numbers yields 16.4 feet. This problem only required two fractions. Others may require longer expressions, but the underlying rule stays the same. When there's a unit on the top of the fraction that's the same as the unit on the bottom, then they cancel each other out. Using this logic and the conversions given above, many units can be converted between and within the different systems.

The conversion between Fahrenheit and Celsius is found in a formula:

$$°C = (°F - 32) \times \frac{5}{9}$$

For example, to convert 78°F to Celsius, the given temperature would be entered into the formula:

$$°C = (78 - 32) \times \frac{5}{9}$$

Solving the equation, the temperature comes out to be 25.56°C. To convert in the other direction, the formula becomes:

$$°F = °C * \frac{9}{5} + 32$$

Remember the order of operations when calculating these conversions.

Solving Problems Involving Elapsed Time, Money, Length, Volume, and Mass

To solve problems, follow these steps: Identify the variables that are known, decide which equation should be used, substitute the numbers, and solve. To solve an equation for the amount of time that has elapsed since an event, use the equation $T = L - E$ where T represents the elapsed time, L represents the later time, and E represents the earlier time.

For example, the Minnesota Vikings have not appeared in the Super Bowl since 1976. If the year is now 2017, how long has it been since the Vikings were in the Super Bowl? The later time, L, is 2017, E = 1976 and the unknown is T. Substituting these numbers, the equation is T = 2017 − 1976, and so T = 41. It has been 41 years since the Vikings have appeared in the Super Bowl. Questions involving total cost can be solved using the formula, $C = I + T$ where C represents the total cost, I represents the cost of the item purchased, and T represents the tax amount. To find the length of a rectangle given the area = 32 square inches and width = 8 inches, the formula $A = L \times W$ can be used. Substitute 32 for A and substitute 8 for w, giving the equation 32 = L×8.

This equation is solved by dividing both sides by 8 to find that the length of the rectangle is 4. The formula for volume of a rectangular prism is given by the equation $V = L \times W \times H$. If the length of a rectangular juice box is 4 centimeters, the width is 2 centimeters, and the height is 8 centimeters, what is the volume of this box? Substituting in the formula we find $V = 4 \times 2 \times 8$, so the volume is 64 cubic centimeters. In a similar fashion as those previously shown, the mass of an object can be calculated given the formula, Mass = Density × Volume.

Measuring and Comparing Lengths of Objects Using Standard Tools

Lengths of objects can be measured using tools such as rulers, yard sticks, meter sticks, and tape measures. Typically, a ruler measures 12 inches, or one foot. For this reason, a ruler is normally used to measure lengths smaller than or just slightly more than 12 inches. Rulers may represent centimeters instead of inches. Some rulers have inches on one side and centimeters on the other. Be sure to recognize what units you are measuring in. The standard ruler measurements are divided into units of 1 inch and normally broken down to $\frac{1}{2}$, $\frac{1}{4}$, $\frac{1}{8}$, and even $\frac{1}{16}$ of an inch for more precise measurements. If measuring in centimeters, the centimeter is likely divided into tenths. To measure the size of a picture, for purposes of buying a frame, a ruler is helpful. If the picture is very large, a yardstick, which measures 3 feet and normally is divided into feet and inches, might be useful. Using metric units, the meter stick measures 1 meter and is divided into 100 centimeters. To measure the size of a window in a home, either a yardstick or meter stick would work. To measure the size of a room, though, a tape measure would be the easiest tool to use. Tape measures can measure up to 10 feet, 25 feet, or more depending on the particular tape measure.

Comparing Relative Sizes of U.S. Customary Units and Metric Units

Measuring length in United States customary units is typically done using inches, feet, yards, and miles. When converting among these units, remember that 12 inches = 1 foot, 3 feet = 1 yard, and 5280 feet = 1 mile. Common customary units of weight are ounces and pounds. The conversion needed is 16 ounces = 1 pound. For customary units of volume ounces, cups, pints, quarts, and gallons are typically used. For conversions, use 8 ounces = 1 cup, 2 cups = 1 pint, 2 pints = 1 quart, and 4 quarts = 1 gallon. For measuring lengths in metric units, know that 100 centimeters = 1 meter, and 1000 meters = 1 kilometer. For metric units of measuring weights, grams and kilograms are often used. Know that 1000 grams = 1 kilogram when making conversions. For metric measures of volume, the most common units are milliliters and liters. Remember that 1000 milliliters = 1 liters.

Practice Questions

1. What is $\frac{12}{60}$ converted to a percentage?
   - a. 0.20
   - b. 20%
   - c. 25%
   - d. 12%
   - e. 1.2%

2. Which of the following is the correct decimal form of the fraction $\frac{14}{33}$ rounded to the nearest hundredth place?
   - a. 0.420
   - b. 0.14
   - c. 0.424
   - d. 0.140
   - e. 0.42

3. Which of the following represents the correct sum of $\frac{14}{15}$ and $\frac{2}{5}$?
   - a. $\frac{20}{15}$
   - b. $\frac{4}{3}$
   - c. $\frac{16}{20}$
   - d. $\frac{4}{5}$
   - e. $\frac{16}{15}$

4. What is the product of $\frac{5}{14}$ and $\frac{7}{20}$?
   - a. $\frac{1}{8}$
   - b. $\frac{35}{280}$
   - c. $\frac{12}{34}$
   - d. $\frac{1}{2}$
   - e. $\frac{7}{140}$

5. What is the result of dividing 24 by $\frac{8}{5}$?

    a. $\frac{5}{3}$

    b. $\frac{3}{5}$

    c. $\frac{120}{8}$

    d. 15

    e. $\frac{24}{5}$

6. Subtract $\frac{5}{14}$ from $\frac{5}{24}$. Which of the following is the correct result?

    a. $\frac{25}{168}$

    b. 0

    c. $-\frac{25}{168}$

    d. $\frac{1}{10}$

    e. $-\frac{1}{10}$

7. Which of the following is a correct mathematical statement?

    a. $\frac{1}{3} < -\frac{4}{3}$

    b. $-\frac{1}{3} > \frac{4}{3}$

    c. $\frac{1}{3} > \frac{4}{3}$

    d. $-\frac{1}{3} \geq \frac{4}{3}$

    e. $\frac{1}{3} > -\frac{4}{3}$

8. Which of the following is INCORRECT?

    a. $-\frac{1}{5} < \frac{4}{5}$

    b. $\frac{4}{5} > -\frac{1}{5}$

    c. $-\frac{1}{5} > \frac{4}{5}$

    d. $\frac{1}{5} > -\frac{4}{5}$

    e. $\frac{4}{5} > \frac{1}{5}$

9. What is the solution to the equation $3(x + 2) = 14x - 5$?
   a. $x = 1$
   b. $x = -1$
   c. $x = 0$
   d. All real numbers
   e. No solution

10. What is the solution to the equation $10 - 5x + 2 = 7x + 12 - 12x$?
    a. $x = 12$
    b. $x = 1$
    c. $x = 0$
    d. All real numbers
    e. No solution

11. Which of the following is the result when solving the equation $4(x + 5) + 6 = 2(2x + 3)$?
    a. $x = 6$
    b. $x = 1$
    c. $x = 26$
    d. All real numbers
    e. No solution

12. How many cases of cola can Lexi purchase if each case is $3.50 and she has $40?
    a. 10
    b. 12
    c. 11.4
    d. 11
    e. 12.5

13. Two consecutive integers exist such that the sum of three times the first and two less than the second is equal to 411. What are those integers?
    a. 103 and 104
    b. 104 and 105
    c. 102 and 103
    d. 100 and 101
    e. 101 and 102

14. In a neighborhood, 15 out of 80 of the households have children under the age of 18. What percentage of the households have children under 18?
    a. 0.1875%
    b. 18.75%
    c. 1.875%
    d. 15%
    e. 1.50%

15. Gina took an algebra test last Friday. There were 35 questions, and she answered 60% of them correctly. How many correct answers did she have?

    a. 35

    b. 20

    c. 21

    d. 25

    e. 18

16. Paul took a written driving test, and he got 12 of the questions correct. If he answered 75% of the total questions correctly, how many problems were there in the test?

    a. 25

    b. 15

    c. 20

    d. 18

    e. 16

17. If a car is purchased for $15,395 with a 7.25% sales tax, how much is the total price?

    a. $15,395.07

    b. $16,511.14

    c. $16,411.13

    d. $15,402

    e. $16,113.10

18. A car manufacturer usually makes 15,412 SUVs, 25,815 station wagons, 50,412 sedans, 8,123 trucks, and 18,312 hybrids a month. About how many cars are manufactured each month?

    a. 120,000

    b. 200,000

    c. 300,000

    d. 12,000

    e. 20,000

19. Each year, a family goes to the grocery store every week and spends $105. About how much does the family spend annually on groceries?

    a. $10,000

    b. $50,000

    c. $500

    d. $5,000

    e. $1,200

20. Bindee is having a barbeque on Sunday and needs 12 packets of ketchup for every 5 guests. If 60 guests are coming, how many packets of ketchup should she buy?

    a. 100

    b. 12

    c. 144

    d. 60

    e. 300

21. A grocery store sold 48 bags of apples in one day, and 9 of the bags contained Granny Smith apples. The rest contained Red Delicious apples. What is the ratio of bags of Granny Smith to bags of Red Delicious that were sold?
    a. 48:9
    b. 39:9
    c. 9:48
    d. 9:39
    e. 39:48

22. If Oscar's bank account totaled $4,000 in March and $4,900 in June, what was the rate of change in his bank account total over those three months?
    a. $900 a month
    b. $300 a month
    c. $4,900 a month
    d. $100 a month
    e. $4,000 a month

23. Erin and Katie work at the same ice cream shop. Together, they always work less than 21 hours a week. In a week, if Katie worked two times as many hours as Erin, how many hours could Erin work?
    a. Less than 7 hours
    b. Less than or equal to 7 hours
    c. More than 7 hours
    d. Less than 8 hours
    e. More than 8 hours

24. From the chart below, which two are preferred by more men than women?

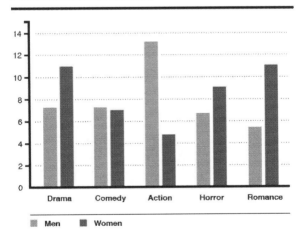

    a. Comedy and Action
    b. Drama and Comedy
    c. Action and Horror
    d. Action and Romance
    e. Romance and Comedy

25. Which type of graph best represents a continuous change over a period of time?
   a. Stacked bar graph
   b. Bar graph
   c. Pie graph
   d. Histogram
   e. Line graph

26. Using the graph below, what is the mean number of visitors for the first 4 hours?

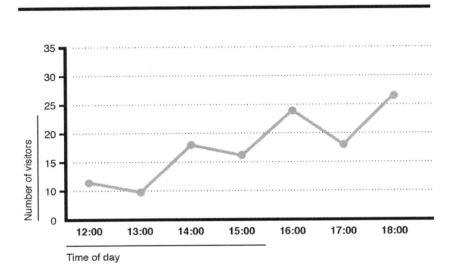

Museum Visitors

Number of visitors

Time of day

   a. 12
   b. 13
   c. 14
   d. 15
   e. 16

27. What is the mode for the grades shown in the chart below?

Science Grades	
Jerry	65
Bill	95
Anna	80
Beth	95
Sara	85
Ben	72
Jordan	98

   a. 65
   b. 33
   c. 95
   d. 90
   e. 84.3

28. What type of relationship is there between age and attention span as represented in the graph below?

Attention Span

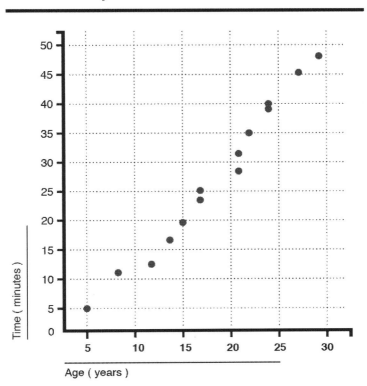

Time (minutes)

Age (years)

a. No correlation
b. Positive correlation
c. Negative correlation
d. Weak correlation
e. Inverse correlation

29. What is the area of the shaded region?

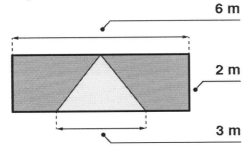

6 m

2 m

3 m

a. 9 m²
b. 12 m²
c. 6 m²
d. 8 m²
e. 4.5 m²

30. What is the volume of the cylinder below?

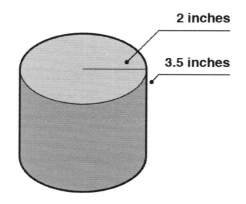

2 inches

3.5 inches

    a. 18.84 in^3
    b. 45.00 in^3
    c. 70.43 in^3
    d. 43.96 in^3
    e. 21.98 in^3

31. How many kiloliters are in 6 liters?
    a. 6,000
    b. 600
    c. 0.006
    d. 0.0006
    e. 0.06

32. How many centimeters are in 3 feet? (Note: 2.54 cm = 1 in)
    a. 0.635
    b. 1.1811
    c. 14.17
    d. 7.62
    e. 91.44

33. Which of the following relations is a function?
    a. {(1, 4), (1, 3), (2, 4), (5, 6)}
    b. {(−1, −1), (−2, −2), (−3, −3), (−4, −4)}
    c. {(0, 0), (1, 0), (2, 0), (1, 1)}
    d. {(1, 0), (1, 2), (1, 3), (1, 4)}
    e. {(−1, 1), (1, −3), (2, 7), (−1, 6)}

34. Find the indicated function value: $f(5)$ for $f(x) = x^2 - 2x + 1$.
    a. 16
    b. 1
    c. 5
    d. 8
    e. Does not exist

35. What is the domain of $f(x) = 4x^2 + 2x - 1$?
    a. $(0, \infty)$
    b. $(-\infty, 0)$
    c. $(-\infty, 4)$
    d. $(-1, 4)$
    e. $(-\infty, \infty)$

36. The function $f(x) = 3.1x + 240$ models the total U.S. population, in millions, x years after the year 1980. Use this function to answer the following question: What is the total U.S. population in 2011? Round to the nearest million.
    a. 336 people
    b. 336 million people
    c. 6,474 people
    d. 647 million people
    e. 64 million people

37. What is the domain of the logarithmic function $f(x) = \log_2(x - 2)$?
    a. 2
    b. $(-\infty, \infty)$
    c. $(0, \infty)$
    d. $(2, \infty)$
    e. $(-\infty, 2)$

38. The function $f(t) = \frac{20,000}{1+10e^{-2t}}$ represents the number of people who catch a disease t weeks after its initial outbreak in a population of 20,000 people. How many people initially had the disease at the time of the initial outbreak? Round to the nearest whole number.
    a. 20,000
    b. 1,818
    c. 2,000
    d. 0
    e. 18,181

39. How is a transposition of a matrix performed?
    a. Multiply each number by negative 1
    b. Switch the rows and columns
    c. Reverse the order of each row
    d. Find the inverse of each number
    e. Divide the first number in each row by the last number in the last column

40. What is the label given to a problem that multiplies a matrix by a constant?
    a. Vector multiplication
    b. Scalar multiplication
    c. Inverse of a matrix
    d. Transposition of a matrix
    e. Product of a matrix

41. What is the range of the polynomial function $f(x) = 2x^2 + 5$?
   a. $(-\infty, \infty)$
   b. $(2, \infty)$
   c. $(0, \infty)$
   d. $(-\infty, 5)$
   e. $[5, \infty)$

42. What are the two values that always describe a vector?
   a. Magnitude and direction
   b. Magnitude and length
   c. Length and position
   d. Direction and position
   e. Magnitude and position

43. For which two values of x are the following functions equal?
$$f(x) = 4x + 4$$
$$g(x) = x^2 + 3x + 2$$

   a. 1, 0
   b. -2, -1
   c. -1, 2
   d. 1, 2
   e. -2, 1

44. The population of coyotes in the local national forest has been declining since 2000. The population can be modeled by the function $y = -(x - 2)^2 + 1600$, where y represents number of coyotes and x represents the number of years past 2000. When will there be no more coyotes?
   a. 2020
   b. 2040
   c. 2012
   d. 2064
   e. 2042

45. Given the linear function $g(x) = \frac{1}{4}x - 2$, which domain value corresponds to a range value of $\frac{1}{8}$?

   a. $\dfrac{17}{2}$

   b. $-\dfrac{63}{32}$

   c. 0

   d. $\dfrac{2}{17}$

   e. $\dfrac{15}{2}$

46. A ball is thrown up from a building that is 800 feet high. Its position s in feet above the ground is given by the function $s = -32t^2 + 90t + 800$, where t is the number of seconds since the ball was thrown. How long will it take for the ball to come back to its starting point? Round your answer to the nearest tenth of a second.

    a. 0 seconds

    b. 2.8 seconds

    c. 3 seconds

    d. 8 seconds

    e. 1.5 seconds

47. What are the zeros of the following cubic function?
$$g(x) = x^3 - 2x^2 - 9x + 18$$

    a. 2, 3

    b. 2, 3, -2

    c. 2, 3, -3,

    d. 2, -2

    e. 0, 2, 3

48. What is the domain of the following rational function?
$$f(x) = \frac{x^3 + 2x + 1}{2 - x}$$

    a. $(-\infty, -2) \cup (-2, \infty)$

    b. $(-\infty, 2) \cup (2, \infty)$

    c. $(2, \infty)$

    d. $(-2, \infty)$

    e. $(-2, 2)$

49. Given the function $f(x) = 4x - 2$, what is the correct form of the simplified difference quotient:
$$\frac{f(x + h) - f(x)}{h}$$

    a. $4x - 1$

    b. $4x$

    c. 4

    d. $4x + h$

    e. $2x - 1$

50. Which set of matrices represents the following system of equations?
$$\begin{cases} x - 2y + 3z = 7 \\ 2x + y + z = 4 \\ -3x + 2y - 2z = -10 \end{cases}$$

a. $\begin{bmatrix} 1 & -2 & 3 \\ 2 & 3 & 1 \\ -3 & 1 & 2 \end{bmatrix}\begin{bmatrix} x \\ y \\ z \end{bmatrix} = \begin{bmatrix} 7 \\ -4 \\ 10 \end{bmatrix}$

b. $\begin{bmatrix} 1 & 2 & -3 \\ -2 & 1 & 2 \\ 3 & 1 & -2 \end{bmatrix}\begin{bmatrix} x \\ y \\ z \end{bmatrix} = \begin{bmatrix} 7 \\ 4 \\ -10 \end{bmatrix}$

c. $\begin{bmatrix} 1 & -2 & 4 \\ -3 & 1 & 7 \\ 2 & 2 & -10 \end{bmatrix}\begin{bmatrix} 3 \\ 1 \\ -2 \end{bmatrix} = \begin{bmatrix} x \\ y \\ z \end{bmatrix}$

d. $\begin{bmatrix} 1 & -2 & 3 \\ 2 & 1 & 1 \\ -3 & 2 & -2 \end{bmatrix}\begin{bmatrix} x \\ y \\ z \end{bmatrix} = \begin{bmatrix} 7 \\ 4 \\ -10 \end{bmatrix}$

e. $\begin{bmatrix} -1 & 2 & -3 \\ 2 & 1 & 1 \\ -3 & 2 & 2 \end{bmatrix}\begin{bmatrix} x \\ y \\ z \end{bmatrix} = \begin{bmatrix} 5 \\ 4 \\ -8 \end{bmatrix}$

51. Which expression is equivalent to $\sqrt[4]{x^6} - \frac{x}{x^3} + x - 2$?

a. $x^{\frac{3}{2}} - x^2 + x - 2$

b. $x^{\frac{2}{3}} - x^{-2} + x - 2$

c. $x^{\frac{3}{2}} - \frac{1}{x^2} + x - 2$

d. $x^{\frac{2}{3}} - \frac{1}{x^2} + x - 2$

e. $x^{\frac{1}{3}} - \frac{1}{x^2} + x - 2$

52. How many possible positive zeros does the polynomial function $f(x) = x^4 - 3x^3 + 2x + x - 3$ have?
   a. 4
   b. 5
   c. 2
   d. 1
   e. 3

53. Which of the following is equivalent to $16^{\frac{1}{4}} \times 16^{\frac{1}{2}}$?
   a. 8
   b. 16
   c. 4
   d. 4,096
   e. 64

54. What is the solution to the following linear inequality?
$$7 - \frac{4}{5}x < \frac{3}{5}$$

    a. $(-\infty, 8)$
    b. $(8, \infty)$
    c. $[8, \infty)$
    d. $(-\infty, 8]$
    e. $(-\infty, \infty)$

55. What is the solution to the following system of linear equations?
$$2x + y = 14$$
$$4x + 2y = -28$$

    a. (0, 0)
    b. (14, -28)
    c. (-14, 28)
    d. All real numbers
    e. There is no solution

56. Triple the difference of five and a number is equal to the sum of that number and 5. What is the number?
    a. 5
    b. 2
    c. 5.5
    d. 2.5
    e. 1

57. Which of the following is perpendicular to the line $4x + 7y = 23$?
    a. $y = -\frac{4}{7}x + 23$
    b. $y = \frac{7}{4}x - 12$
    c. $4x + 7y = 14$
    d. $y = -\frac{7}{4}x + 11$
    e. $y = \frac{4}{7}x - 12$

58. What is the solution to the following system of equations?
$$2x - y = 6$$
$$y = 8x$$

    a. (1, 8)
    b. (-1, -8)
    c. (-1, 8)
    d. All real numbers.
    e. There is no solution.

59. The following set represents the test scores from a university class: {35, 79, 80, 87, 87, 90, 92, 95, 95, 98, 99}. If the outlier is removed from this set, which of the following is TRUE?
   a. The mean and the median will decrease.
   b. The mean and the median will increase.
   c. The mean and the mode will increase.
   d. The mean and the mode will decrease.
   e. The mean, median, and mode will increase.

60. The mass of the moon is about 7.348×10^{22} kilograms and the mass of Earth is 5.972×10^{24} kilograms. How many times greater is Earth's mass than the moon's mass?
   a. 8.127×10^1
   b. 8.127
   c. 812.7
   d. 8.127×10^{-1}
   e. 0.8127

61. What is the equation of the line that passes through the two points (-3, 7) and (-1, -5)?
   a. $y = 6x + 11$
   b. $y = 6x$
   c. $y = -6x - 11$
   d. $y = -6x$
   e. $y = 6x - 11$

62. The percentage of smokers above the age of 18 in 2000 was 23.2 percent. The percentage of smokers above the age of 18 in 2015 was 15.1 percent. Find the average rate of change in the percent of smokers above the age of 18 from 2000 to 2015.
   a. −0.54 percent
   b. −54 percent
   c. −5.4 percent
   d. −0.46 percent
   e. −46 percent

63. A study of adult drivers finds that it is likely that an adult driver wears his seatbelt. Which of the following could be the probability that an adult driver wears his seat belt?
   a. 0.90
   b. 0.05
   c. 0.25
   d. 0
   e. 1.5

64. In order to estimate deer population in a forest, biologists obtained a sample of deer in that forest and tagged each one of them. The sample had 300 deer in total. They returned a week later and harmlessly captured 400 deer, and found 5 were tagged. Use this information to estimate how many total deer were in the forest.
   a. 24,000 deer
   b. 30,000 deer
   c. 40,000 deer
   d. 100,000 deer
   e. 120,000 deer

65. Which of the following is the equation of a vertical line that runs through the point (1, 4)?

    a. $x = 1$

    b. $y = 1$

    c. $x = 4$

    d. $y = 4$

    e. $x = y$

66. What is the missing length x?

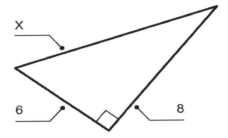

    a. 6

    b. −10

    c. 10

    d. 100

    e. 14

67. What is the correct factorization of the following binomial?
$$2y^3 - 128$$

    a. $2(y + 8)(y - 8)$

    b. $2(y + 4)(y^2 - 4y)$

    c. $2(y + 4)(y - 4)^2$

    d. $2(y - 4)^3$

    e. $2(y - 4)(y^2 + 4y + 16)$

68. What is the simplified form of $(4y^3)^4(3y^7)^2$?

    a. $12y^{26}$

    b. $2{,}304y^{16}$

    c. $12y^{14}$

    d. $2{,}304y^{26}$

    e. $12y^{16}$

69. Use the graph below entitled "Projected Temperatures for Tomorrow's Winter Storm" to answer the question.

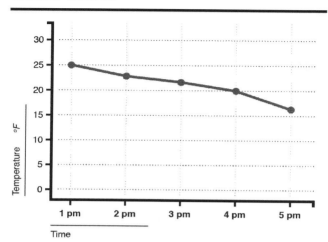

Projected Temperatures for Tomorrow's Winter Storm

What is the expected temperature at 3:00 p.m.?
   a. 25 degrees
   b. 22 degrees
   c. 20 degrees
   d. 16 degrees
   e. 18 degrees

70. The number of members of the House of Representatives varies directly with the total population in a state. If the state of New York has 19,800,000 residents and has 27 total representatives, how many should Ohio have with a population of 11,800,000?
   a. 10
   b. 16
   c. 11
   d. 5
   e. 12

71. Which of the statements below is a statistical question?
   a. What was your grade on the last test?
   b. What were the grades of the students in your class on the last test?
   c. What kind of car do you drive?
   d. What was Sam's time in the marathon?
   e. What textbooks does Marty use this semester?

Use the following information to answer the next three questions, rounding to the closest minute. Eva Jane is practicing for an upcoming 5K run. She has recorded the following times (in minutes):

25, 18, 23, 28, 30, 22.5, 23, 33, 20

72. What is Eva Jane's mean time?
    a. 26 minutes
    b. 19 minutes
    c. 24.5 minutes
    d. 23 minutes
    e. 25 minutes

73. What is the mode of Eva Jane's times?
    a. 16 minutes
    b. 20 minutes
    c. 23 minutes
    d. 33 minutes
    e. 25 minutes

74. What is Eva Jane's median time?
    a. 23 minutes
    b. 17 minutes
    c. 28 minutes
    d. 19 minutes
    e. 25 minutes

75. What is the area of the following figure?

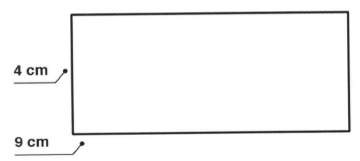

    a. 26 cm
    b. 36 cm
    c. 13 cm^2
    d. 36 cm^2
    e. 65 cm^2

76. What is the volume of the given figure?

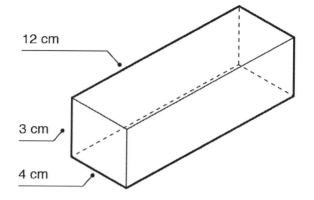

12 cm

3 cm

4 cm

a. 36 cm²
b. 144 cm³
c. 72 cm³
d. 36 cm³
e. 144 cm²

77. What type of units are used to describe surface area?
a. Square
b. Cubic
c. Single
d. Quartic
e. Volumetric

78. What is the perimeter of the following figure?

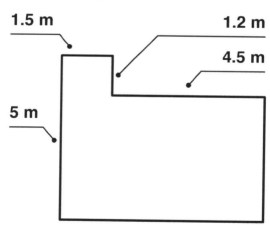

1.5 m   1.2 m

4.5 m

5 m

a. 13.4 m
b. 22 m
c. 12.2 m
d. 22.5 m
e. 24.4 m

79. Which equation correctly shows how to find the surface area of a cylinder?

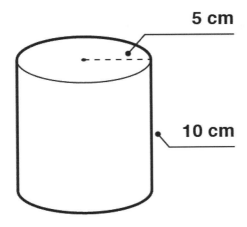

5 cm

10 cm

a. $SA = 2\pi \times 5 \times 10 + 2(\pi 5^2)$
b. $SA = 5 \times 2\pi \times 5$
c. $SA = 2\pi 5^2$
d. $SA = 2\pi \times 10 + \pi 5^2$
e. $SA = 2\pi \times 5 \times 10 + \pi 5^2$

80. Which shapes could NOT be used to compose a hexagon?

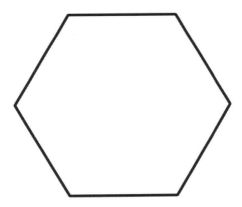

a. Six triangles
b. One rectangle and two triangles
c. Two rectangles
d. Two trapezoids
e. One rectangle and four triangles

Answer Explanations

1. B: The fraction $\frac{12}{60}$ can be reduced to $\frac{1}{5}$, in lowest terms. First, it must be converted to a decimal. Dividing 1 by 5 results in 0.2. Then, to convert to a percentage, move the decimal point two units to the right and add the percentage symbol. The result is 20%.

2. E: If a calculator is used, divide 33 into 14 and keep two decimal places. If a calculator is not used, multiply both the numerator and denominator times 3. This results in the fraction $\frac{42}{99}$, and hence a decimal of 0.42.

3. B: Common denominators must be used. The LCD is 15, and:

$$\frac{2}{5} = \frac{6}{15}$$

Therefore:

$$\frac{14}{15} + \frac{6}{15} = \frac{20}{15}$$

and in lowest terms the answer is $\frac{4}{3}$. A common factor of 5 was divided out of both the numerator and denominator.

4. A: A product is found by multiplication. Multiplying two fractions together is easier when common factors are cancelled first to avoid working with larger numbers.

$$\frac{5}{14} \times \frac{7}{20} = \frac{5}{2 \times 7} \times \frac{7}{5 \times 4}$$

$$\frac{1}{2} \times \frac{1}{4} = \frac{1}{8}$$

5. D: Division is completed by multiplying times the reciprocal. Therefore:

$$24 \div \frac{8}{5} = \frac{24}{1} \times \frac{5}{8}$$

$$\frac{3 \times 8}{1} \times \frac{5}{8} = \frac{15}{1} = 15$$

6. C: Common denominators must be used. The LCD is 168, so each fraction must be converted to have 168 as the denominator.

$$\frac{5}{24} - \frac{5}{14} = \frac{5}{24} \times \frac{7}{7} - \frac{5}{14} \times \frac{12}{12}$$

$$\frac{35}{168} - \frac{60}{168} = -\frac{25}{168}$$

7. E: The correct mathematical statement is the one in which the number to the left on the number line is less than the number to the right on the number line. It is written in Choice *E* that:

$$\frac{1}{3} > -\frac{4}{3}$$

which is the same as:

$-\frac{4}{3} < \frac{1}{3}$, a correct statement.

8. C: $-\frac{1}{5} > \frac{4}{5}$ is an incorrect statement. The expression on the left is negative, which means that it is smaller than the expression on the right. As it is written, the inequality states that the expression on the left is greater than the expression on the right, which is not true.

9. A: First, the distributive property must be used on the left side. This results in:

$$3x + 6 = 14x - 5$$

The addition property is then used to add 5 to both sides, and then to subtract $3x$ from both sides, resulting in $11 = 11x$. Finally, the multiplication property is used to divide each side by 11. Therefore, $x = 1$ is the solution.

10. D: First, like terms are collected to obtain:

$$12 - 5x = -5x + 12$$

Then, if the addition principle is used to move the terms with the variable, $5x$ is added to both sides and the mathematical statement $12 = 12$ is obtained. This is always true; therefore, all real numbers satisfy the original equation.

11. E: The distributive property is used on both sides to obtain:

$$4x + 20 + 6 = 4x + 6$$

Then, like terms are collected on the left, resulting in:

$$4x + 26 = 4x + 6$$

Next, the addition principle is used to subtract $4x$ from both sides, and this results in the false statement $26 = 6$. Therefore, there is no solution.

12. D: This is a one-step real-world application problem. The unknown quantity is the number of cases of cola to be purchased. Let x be equal to this amount. Because each case costs $3.50, the total number of cases multiplied by $3.50 must equal $40. This translates to the mathematical equation:

$$3.5x = 40$$

Divide both sides by 3.5 to obtain $x = 11.4286$, which has been rounded to four decimal places. Because cases are sold whole, and there is not enough money to purchase 12 cases, 11 cases is the correct answer.

13. A: First, the variables have to be defined. Let x be the first integer; therefore, $x + 1$ is the second integer. This is a two-step problem. The sum of three times the first and two less than the second is translated into the following expression:

$$3x + (x + 1 - 2)$$

This expression is set equal to 411 to obtain:

$$3x + (x + 1 - 2) = 411$$

The left-hand side is simplified to obtain:

$$4x - 1 = 411$$

The addition and multiplication properties are used to solve for x. First, add 1 to both sides and then divide both sides by 4 to obtain $x = 103$. The next consecutive integer is 104.

14. B: First, the information is translated into the ratio $\frac{15}{80}$. To find the percentage, translate this fraction into a decimal by dividing 15 by 80. The corresponding decimal is 0.1875. Move the decimal point two units to the right to obtain the percentage 18.75%.

15. C: Gina answered 60% of 35 questions correctly; 60% can be expressed as the decimal 0.60. Therefore, she answered $0.60 \times 35 = 21$ questions correctly.

16. E: The unknown quantity is the number of total questions on the test. Let x be equal to this unknown quantity. Therefore, $0.75x = 12$. Divide both sides by 0.75 to obtain $x = 16$.

17. B: If sales tax is 7.25%, the price of the car must be multiplied times 1.0725 to account for the additional sales tax. Therefore:

$$15{,}395 \times 1.0725 = 16{,}511.1375$$

This amount is rounded to the nearest cent, which is $16,511.14.

18. A: Rounding can be used to find the best approximation. All of the values can be rounded to the nearest thousand. 15,412 SUVs can be rounded to 15,000. 25,815 station wagons can be rounded to 26,000. 50,412 sedans can be rounded to 50,000. 8,123 trucks can be rounded to 8,000. Finally, 18,312 hybrids can be rounded to 18,000. The sum of the rounded values is 117,000, which is closest to 120,000.

19. D: There are 52 weeks in a year, and if the family spends $105 each week, that amount is close to $100. A good approximation is $100 a week for 50 weeks, which is found through the product $50 \times 100 = \$5{,}000$.

20. C: This problem involves ratios and percentages. If 12 packets are needed for every 5 people, this statement is equivalent to the ratio $\frac{12}{5}$. The unknown amount x is the number of ketchup packets needed for 60 people. The proportion:

$$\frac{12}{5} = \frac{x}{60}$$

must be solved. Cross-multiply to obtain:

$$12 \times 60 = 5x$$

Therefore, $720 = 5x$. Divide each side by 5 to obtain $x = 144$.

21. D: There were 48 total bags of apples sold. If 9 bags were Granny Smith and the rest were Red Delicious, then $48 - 9 = 39$ bags were Red Delicious. Therefore, the ratio of Granny Smith to Red Delicious is 9:39.

22. B: The average rate of change is found by calculating the difference in dollars over the elapsed time. Therefore, the rate of change is equal to:

$$(\$4,900 - \$4,000) \div 3 \text{ months}$$

which is equal to $\$900 \div 3$ or $\$300$ per month.

23. A: Let x be the unknown, the number of hours Erin can work. We know Katie works $2x$, and the sum of all hours is less than 21. Therefore:

$$x + 2x < 21$$

which simplifies into $3x < 21$. Solving this results in the inequality $x < 7$ after dividing both sides by 3. Therefore, Erin can work less than 7 hours.

24. A: The chart is a bar chart showing how many men and women prefer each genre of movies. The dark gray bars represent the number of women, while the light gray bars represent the number of men. The light gray bars are higher and represent more men than women for the genres of Comedy and Action.

25. E: A line graph represents continuous change over time. The line on the graph is continuous and not broken, as on a scatter plot. Stacked bar graphs are used when comparing multiple variables at one time. They combine some elements of both pie charts and bar graphs, using the organization of bar graphs and the proportionality aspect of pie charts. A bar graph may show change but isn't necessarily continuous over time. A pie graph is better for representing percentages of a whole. Histograms are best used in grouping sets of data in bins to show the frequency of a certain variable.

26. C: The mean for the number of visitors during the first 4 hours is 14. The mean is found by calculating the average for the four hours. Adding up the total number of visitors during those hours gives:

$$12 + 10 + 18 + 16 = 56$$

Dividing total visitors by four hours gives average visitors per hour, $56 \div 4 = 14$.

27. C: The mode for a set of data is the value that occurs the most. The grade that appears the most is 95. It's the only value that repeats in the set. The mean is around 84.3.

28. B: The relationship between age and time for attention span is a positive correlation because the general trend for the data is up and to the right. As the age increases, so does attention span.

29. A: The area of the shaded region is calculated in a few steps. First, the area of the rectangle is found using the formula:

$$A = length \times width = 6 \text{ m} \times 2 \text{ m} = 12 \text{ m}^2$$

Second, the area of the triangle is found using the formula:

$$A = \frac{1}{2} \times base \times height = \frac{1}{2} \times 3 \text{ m} \times 2 \text{ m} = 3 \text{ m}^2$$

The last step is to take the rectangle area and subtract the triangle area. The area of the shaded region is:

$$A = 12 \text{ m}^2 - 3 \text{ m}^2 = 9 \text{ m}^2$$

30. D: The volume for a cylinder is found by using the formula:

$$V = \pi r^2 h = \pi (2 \text{ in})^2 \times 3.5 \text{ in} = 43.96 \text{ in}^3$$

31. C: There are 0.006 kiloliters in 6 liters because 1 liter is 0.001 kiloliters. The conversion comes from the metric prefix -kilo which has a value of 1000. Thus, 1 kiloliter is 1000 liters, and 1 liter is 0.001 kiloliters.

32. E: The conversion between feet and centimeters requires a middle term. As there are 2.54 centimeters in 1 inch, the conversion between inches and feet must be found. As there are 12 inches in a foot, the fractions can be set up as follows:

$$3 \text{ ft} \times \frac{12 \text{ in}}{1 \text{ ft}} \times \frac{2.54 \text{ cm}}{1 \text{ in}}$$

The feet and inches cancel out to leave only centimeters for the answer. The numbers are calculated across the top and bottom to yield:

$$\frac{3 \times 12 \times 2.54}{1 \times 1} = 91.44$$

The number and units used together form the answer of 91.44 cm.

33. B: The only relation in which every x-value corresponds to exactly one y-value is the relation given in B, making it a function. The other relations have the same x-value paired up to different y-values, which goes against the definition of functions.

34. A: To find a function value, plug in the number given for the variable and evaluate the expression, using the order of operations (parentheses, exponents, multiplication, division, addition, subtraction). The function given is a polynomial function:

$$f(5) = 5^2 - 2 \times 5 + 1 = 25 - 10 + 1 = 16$$

35. E: The function given is a polynomial function. Anything can be plugged into a polynomial function to get an output. Therefore, its domain is all real numbers, which is expressed in interval notation as $(-\infty, \infty)$.

36. B: The variable x represents the number of years after 1980. The year 2011 was 31 years after 1980, so plug 31 into the function to obtain:

$$f(31) = 3.1 \times 31 + 240 = 336.1$$

This value rounds to 336 and represents 336 million people.

37. D: The argument of a logarithmic function has to be greater than or equal to zero. Basically, one cannot take the logarithm of a negative number or 0. Therefore, to find the domain, set the argument greater than 0 and solve the inequality. This results in $x - 2 > 0$, or $x > 2$.

Therefore, in order to obtain an output of the function, the number plugged into the function must be greater than 2. This domain is represented as $(2, \infty)$.

38. B: The time of the initial outbreak corresponds to $t = 0$. Therefore, 0 must be plugged into the function. This results in:

$$\frac{20,000}{1 + 10e^0} = \frac{20,000}{1 + 10} = \frac{20,000}{11} = 1,818.182$$

which rounds to 1,818. Therefore, there were 1,818 people in the population that initially had the disease.

39. B: The correct choice is B because the definition of transposing a matrix says that the rows and columns should be switched. For example, consider the following matrix:

$$\begin{bmatrix} 3 & 4 \\ 2 & 5 \\ 1 & 6 \end{bmatrix}$$

This can be transposed into:

$$\begin{bmatrix} 3 & 2 & 1 \\ 4 & 5 & 6 \end{bmatrix}$$

Notice that the first row, 3 and 4, becomes the first column. The second row, 2 and 5, becomes the second column. This is an example of transposing a matrix.

40. B: The correct answer is Choice B because multiplying a matrix by a constant is called scalar multiplication. A scalar is a constant number, which means the only thing it changes about a matrix is its magnitude. For a given matrix, $\begin{bmatrix} 3 & 4 \\ 6 & 5 \end{bmatrix}$, scalar multiplication can be applied by multiplying by 2, which yields the matrix $\begin{bmatrix} 6 & 8 \\ 12 & 10 \end{bmatrix}$.

Notice that the dimensions of the matrix did not change, just the magnitude of the numbers.

41. E: This is a parabola that opens up, as the coefficient on the x^2 term is positive. The smallest number in its range occurs when plugging 0 into the function $f(0) = 5$. Any other output is a number larger than 5, even when a positive number is plugged in. When a negative number gets plugged into the function, the output is positive, and same with a positive number. Therefore, the domain is written as $[5, \infty)$ in interval notation.

42. A: The vector is described as having both magnitude and direction. The magnitude is the size of the vector and the direction is the path along with which the force is being applied. The second answer choice has magnitude and length, which are essentially the same. The third, fourth, and fifth answer choices include length and position, but position is not part of the description of a vector.

43. C: First, set the functions equal to one another, resulting in:

$$x^2 + 3x + 2 = 4x + 4$$

This is a quadratic equation, so the equivalent equation in standard form is:

$$x^2 - x - 2 = 0$$

This equation can be solved by factoring into:

$$(x - 2)(x + 1) = 0$$

Setting both factors equal to zero results in $x = 2$ and $x = -1$.

44. E: There will be no more coyotes when the population is 0, so set y equal to 0 and solve the quadratic equation:

$$0 = -(x - 2)^2 + 1600$$

Subtract 1600 from both sides, and divide through by -1. This results in:

$$1600 = (x - 2)^2$$

Then, take the square root of both sides. This process results in the following equation: $\pm 40 = x - 2$. Adding 2 to both sides results in two solutions: $x = 42$ and $x = -38$.

Because the problem involves years after 2000, the only solution that makes sense is 42. Add 42 to 2000, so therefore in 2042 there will be no more coyotes.

45. A: The range value is given, and this is the output of the function. Therefore, the function must be set equal to $\frac{1}{8}$ and solved for x. Thus:

$$\frac{1}{8} = \frac{1}{4}x - 2$$

needs to be solved. The fractions can be cleared by multiplying times the LCD 8. This results in:

$$1 = 2x - 16$$

Add 16 to both sides and divide by 2 to obtain:

$$x = \frac{17}{2}$$

46. B: The ball is back at the starting point when the function is equal to 800 feet. Therefore, this results in solving the equation:

$$800 = -32t^2 + 90t + 800$$

Subtract 800 off of both sides and factor the remaining terms to obtain:

$$0 = 2t(-16t + 45)$$

Setting both factors equal to 0 results in $t = 0$, which is when the ball was thrown up initially, and:

$$t = \frac{45}{16} = 2.8 \text{ seconds}$$

Therefore, it will take the ball 2.8 seconds to come back down to its starting point.

47. C: To find the zeros, set the function equal to 0 and factor the polynomial. Because there are four terms, it should be factored by grouping. Factor a common factor out of the first set of two terms, and then find a shared binomial factor in the second set of two terms. This results in:

$$x^2(x - 2) - 9(x - 2) = 0$$

The binomial can then be factored out of each set to get:

$$(x^2 - 9)(x - 2) = 0$$

This can be factored further as:

$$(x + 3)(x - 3)(x - 2) = 0$$

Setting each factor equal to zero and solving results in the three zeros −3, 3, and 2.

48. B: Given a rational function, the expression in the denominator can never be equal to 0. To find the domain, set the denominator equal to 0 and solve for x. This results in $2 - x = 0$, and its solution is $x = 2$. This value needs to be excluded from the set of all real numbers, and therefore the domain written in interval notation is $(-\infty, 2) \cup (2, \infty)$.

49. C: Plugging the function into the formula results in:

$$\frac{4(x + h) - 2 - (4x - 2)}{h}$$

which is simplified to:

$$\frac{4x + 4h - 2 - 4x + 2}{h} = \frac{4h}{h} = 4$$

This value is also equal to the derivative of the given function. The derivative of a linear function is its slope.

50. D: The correct matrix to describe the given system of equations is the Choice *D* because it has values that correspond to the coefficients in the right order. The top row corresponds to the coefficient in the first equation, the second row corresponds to the coefficients in the second equation, and the third row corresponds to the coefficients in the third equation. The second matrix (Choice *B*) is filled with the three variables in the system. One thing to also look for is the sign on the numbers, to make sure the signs are correct from the equation to the matrix.

51. C: By switching from a radical expression to rational exponents:

$$\sqrt[4]{x^6} = x^{\frac{6}{4}} = x^{\frac{3}{2}}$$

Also, properties of exponents can be used to simplify $\frac{x}{x^3}$ into:

$$x^{1-3} = x^{-2} = \frac{1}{x^2}$$

The other terms can be left alone, resulting in an equivalent expression:

$$x^{\frac{3}{2}} - \frac{1}{x^2} + x - 2$$

52. E: Using Descartes' Rule of Signs, count the number of sign changes in coefficients in the polynomial. This results in the number of possible positive zeros. The coefficients are 1, −3, 2, 1, and −3, so the sign changes from 1 to −3, −3 to 2, and 1 to −3, a total of 3 times. Therefore, there are at most 3 positive zeros.

53. A: The first step is to simplify the expression. The second term, $16^{\frac{1}{2}}$, can be rewritten as $\sqrt{16}$, since fractional exponents represent roots. This can then be simplified to 4, because 16 is the perfect square of 4. The first term can be evaluated using the power rule for exponents: an exponent to a second exponent is multiplied by that second exponent. The first term can thus be rewritten as:

$$16^{\frac{1}{4}} = \left(16^{\frac{1}{2}}\right)^{\frac{1}{2}} = 4^{\frac{1}{2}} = 2$$

Combining these terms, the expression is evaluated as follows:

$$16^{\frac{1}{4}} \times 16^{\frac{1}{2}} = 2 \times 4 = 8$$

54. B: The goal is to first isolate the variable. The fractions can easily be cleared by multiplying the entire inequality by 5, resulting in $35 - 4x < 3$.

Then, subtract 35 from both sides and divide by −4. This results in $x > 8$.

Notice the inequality symbol has been flipped because both sides were divided by a negative number. The solution set, all real numbers greater than 8, is written in interval notation as $(8, \infty)$. A parenthesis shows that 8 is not included in the solution set.

55. E: This system can be solved using the method of substitution. Solving the first equation for y results in $y = 14 - 2x$.

Plugging this into the second equation gives:

$$4x + 2(14 - 2x) = -28$$

which simplifies to $28 = -28$, an untrue statement. Therefore, this system has no solution because no x-value will satisfy the system.

56. D: Let x be the unknown number. The difference indicates subtraction, and sum represents addition. To triple the difference, it is multiplied by 3. The problem can be expressed as the following equation:

$$3(5 - x) = x + 5$$

Distributing the 3 results in:

$$15 - 3x = x + 5$$

Subtract 5 from both sides, add $3x$ to both sides, and then divide both sides by 4. This results in:

$$x = \frac{10}{4} = \frac{5}{2} = 2.5$$

57. B: The slopes of perpendicular lines are negative reciprocals, meaning their product is equal to -1. The slope of the line given needs to be found. Its equivalent form in slope-intercept form is:

$$y = -\frac{4}{7}x + \frac{23}{7}$$

so its slope is $-\frac{4}{7}$. The negative reciprocal of this number is $\frac{7}{4}$. The only line in the options given with this same slope is $y = \frac{7}{4}x - 12$.

58. B: This system can be solved using substitution. Plug the second equation in for y in the first equation to obtain:

$$2x - 8x = 6$$

which simplifies to $-6x = 6$.

Divide both sides by 6 to get $x = -1$, which is then back-substituted into either original equation to obtain $y = -8$.

59. B: The outlier is 35. When a small outlier is removed from a data set, the mean and the median increase. The first step in this process is to identify the outlier, which is the number that lies away from the given set. Once the outlier is identified, the mean and median can be recalculated.

The mean will be affected because it averages all of the numbers. The median will be affected because it finds the middle number, which is subject to change because a number is lost. The mode will most likely not change because it is the number that occurs the most, which will not be the outlier if there is only one outlier.

60. A: Division can be used to solve this problem. The division necessary is:

$$\frac{5.972 \times 10^{24}}{7.348 \times 10^{22}}$$

To compute this division, divide the constants first then use algebraic laws of exponents to divide the exponential expression.

This results in about 0.8127×10^2, which written in scientific notation is 8.127×10^1.

61. C: First, the slope of the line must be found. This is equal to the change in y over the change in x, given the two points. Therefore, the slope is -6. The slope and one of the points are then plugged into the slope-intercept form of a line:

$$y - y_1 = m(x - x_1)$$

This results in:

$$y - 7 = -6(x + 3)$$

The -6 is distributed and the equation is solved for y to obtain $y = -6x - 11$.

62. A: The formula for the rate of change is the same as slope: change in y over change in x. The y-value in this case is percentage of smokers and the x-value is year. The change in percentage of smokers from 2000 to 2015 was 8.1 percent.

The change in x was $2000 - 2015 = -15$. Therefore:

$$\frac{8.1\%}{-15} = -0.54\%$$

The percentage of smokers decreased 0.54 percent each year.

63. A: The probability of 0.9 is closer to 1 than any of the other answers. The closer a probability is to 1, the greater the likelihood that the event will occur. The probability of 0.05 shows that it is very unlikely that an adult driver will wear their seatbelt because it is close to zero. A zero probability means that it will not occur. The probability of 0.25 is closer to zero than to one, so it shows that it is unlikely an adult will wear their seatbelt. Choice E is wrong because probability must fall between 0 and 1.

64. A: A proportion should be used to solve this problem. The ratio of tagged to total deer in each instance is set equal, and the unknown quantity is a variable x. The proportion is:

$$\frac{300}{x} = \frac{5}{400}$$

Cross-multiplying gives $120,000 = 5x$, and dividing through by 5 results in 24,000.

65. A: A vertical line has the same x value for any point on the line. Other points on the line would be (1, 3), (1, 5), (1, 9), etc. Mathematically, this is written as $x = 1$. A vertical line is always of the form $x = a$ for some constant a.

66. C: The Pythagorean Theorem can be used to find the missing length x because it is a right triangle. The theorem states that:

$$6^2 + 8^2 = x^2$$

which simplifies into $100 = x^2$. Taking the positive square root of both sides results in the missing value $x = 10$.

67. E: First, the common factor 2 can be factored out of both terms, resulting in:

$$2(y^3 - 64)$$

The resulting binomial is a difference of cubes that can be factored using the rule:

$$a^3 - b^3 = (a - b)(a^2 + ab + b^2)$$

with $a = y$ and $b = 4$. Therefore, the result is:

$$2(y - 4)(y^2 + 4y + 16)$$

68. D: The exponential rules $(ab)^m = a^m b^m$ and $(a^m)^n = a^{mn}$ can be used to rewrite the expression as:

$$4^4 y^{12} \times 3^2 y^{14}$$

The coefficients are multiplied together and the exponential rule:

$$a^m a^n = a^{m+n}$$

is then used to obtain the simplified form $2{,}304 y^{26}$.

69. B: Look on the horizontal axis to find 3:00 p.m. Move up from 3:00 p.m. to reach the dot on the graph. Move horizontally to the left to the horizontal axis to between 20 and 25; the best answer choice is 22. The answer of 25 is too high above the projected time on the graph, and the answers of 20, 16, and 18 degrees are too low.

70. B: The number of representatives varies directly with the population, so the equation necessary is $N = k \times P$, where N is number of representatives, k is the variation constant, and P is total population in millions.

Plugging in the information for New York allows k to be solved for. This process gives:

$$27 = k \times 19.8$$

so $k = 1.36$. Therefore, the formula for number of representatives given total population in millions is:

$$N = 1.36 \times P$$

Plugging in $P = 11.8$ for Ohio results in $N = 16.05$, which rounds to 16 total representatives.

71. B: This is a statistical question because to determine this answer one would need to collect data from each person in the class and it is expected the answers would vary. The other answers do not require data to be collected from multiple sources, therefore the answers will not vary.

72. E: The mean is found by adding all the times together and dividing by the number of times recorded:

$$25 + 18 + 23 + 28 + 30 + 22.5 + 23 + 33 + 20 = 222.5$$

divided by $9 = 24.7$

Rounding to the nearest minute, the mean is 25.

73. C: The mode is the time from the data set that occurs most often. The number 23 occurs twice in the data set, while all others occur only once, so the mode is 23.

74. A: To find the median of a data set, you must first list the numbers from smallest to largest, and then find the number in the middle. If there are two numbers in the middle, add the two numbers in the middle together and divide by 2. Putting this list in order from smallest to greatest yields 18, 20, 22.5, 23, 23, 25, 28, 30, and 33, where 23 is the middle number, so 23 minutes is the median.

75. D: The area for a rectangle is found by multiplying the length by the width. The area is also measured in square units, so the correct answer is Choice *D*. The answer of 26 is the perimeter. The answer of 13 is found by adding the two dimensions instead of multiplying.

76. B: The volume of a rectangular prism is found by multiplying the length by the width by the height. This formula yields an answer of 144 cubic units. The answer must be in cubic units because volume involves all three dimensions. Each of the other answers have only two dimensions that are multiplied, and one dimension is forgotten, as in *D*, where 12 and 3 are multiplied, or have incorrect units, as in *E*.

77. A: Surface area is a type of area, which means it is measured in square units. Cubic units are used to describe volume, which has three dimensions multiplied by one another. Quartic units describe measurements multiplied in four dimensions.

78. B: The perimeter is found by adding the length of all the exterior sides. When the given dimensions are added, the perimeter is 22 meters. The equation to find the perimeter can be:

$$P = 5 + 1.5 + 1.2 + 4.5 + 3.8 + 6 = 22$$

The last two dimensions can be found by subtracting 1.2 from 5, and adding 1.5 and 4.5, respectively.

79. A: The surface area for a cylinder is the sum of the two circle bases and the rectangle formed on the side. This is easily seen in the net of a cylinder.

The Net of a Cylinder

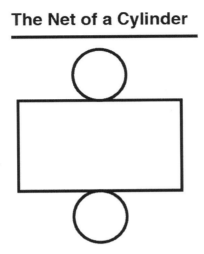

The area of a circle is found by multiplying pi times the radius squared. The rectangle's area is found by multiplying the circumference of the circle by the height. The equation:

$$SA = 2\pi \times 5 \times 10 + 2(\pi 5^2)$$

shows the area of the rectangle as $2\pi \times 5 \times 10$, which yields 314. The area of the bases is found by $\pi 5^2$, which yields 78.5, and then is multiplied by 2 for the two bases. Adding these together gives 471.

80. C: A hexagon can be formed by any combination of the given shapes except for two rectangles. There are no two rectangles that can make up a hexagon.

ACT Reading

Key Ideas and Details

Central Ideas and Themes

The **topic** of a text is the general subject matter. Text topics can usually be expressed in one word, or a few words at most. Additionally, readers should ask themselves what point the author is trying to make. This point is the **main idea** of the text, the one thing the author wants readers to know concerning the topic. Once the author has established the main idea, he or she will support the main idea by supporting details. **Supporting details** are evidence that support the main idea and include personal testimonies, examples, or statistics.

One analogy for these components and their relationships is that a text is like a well-designed house. The topic is the roof, covering all rooms. The main idea is the frame. The supporting details are the various rooms. To identify the topic of a text, readers can ask themselves what or who the author is writing about in the paragraph. To locate the main idea, readers can ask themselves what one idea the author wants readers to know about the topic. To identify supporting details, readers can put the main idea into question form and ask, "what does the author use to prove or explain their main idea?"

Let's look at an example. An author is writing an essay about the Amazon rainforest and trying to convince the audience that more funding should go into protecting the area from deforestation. The author makes the argument stronger by including evidence of the benefits of the rainforest: it provides habitats to a variety of species, it provides much of the earth's oxygen which in turn cleans the atmosphere, and it is the home to medicinal plants that may be the answer to some of the world's deadliest diseases. Here is an outline of the essay looking at topic, main idea, and supporting details:

- Topic: Amazon rainforest
- Main Idea: The Amazon rainforest should receive more funding to protect it from deforestation.
- Supporting Details:
    - 1. It provides habitats to a variety of species
    - 2. It provides much of the earth's oxygen which in turn cleans the atmosphere
    - 3. It is home to medicinal plants that may be the answer to some of the deadliest diseases.

Notice that the topic of the essay is listed in a few key words: "Amazon rainforest." The main idea tells us what about the topic is important: that the topic should be funded in order to prevent deforestation. Finally, the supporting details are what author relies on to convince the audience to act or to believe in the truth of the main idea.

Summarizing Information and Ideas

An important skill is the ability to read a complex text and then reduce its length and complexity by focusing on the key events and details. A **summary** is a shortened version of the original text, written by the reader in their own words. The summary should be shorter than the original text, and it must be thoughtfully formed to include critical points from the original text.

In order to effectively summarize a complex text, it's necessary to understand the original source and identify the major points covered. It may be helpful to outline the original text to get the big picture and avoid getting bogged down in the minor details. For example, a summary wouldn't include a statistic from the original source unless it was the major focus of the text. It's also important for readers to use their own words, yet retain the original meaning of the passage. The key to a good summary is emphasizing the main idea without changing the focus of the original information.

The more complex a text, the more difficult it can be to summarize. Readers must evaluate all points from the original source and then filter out what they feel are the less necessary details. Only the essential ideas should remain. The summary often mirrors the original text's organizational structure. For example, in a problem-solution text structure, the author typically presents readers with a problem and then develops solutions through the course of the text. An effective summary would likely retain this general structure, rephrasing the problem and then reporting the most useful or plausible solutions.

Paraphrasing is somewhat similar to summarizing. It calls for the reader to take a small part of the passage and list or describe its main points. Paraphrasing is more than rewording the original passage, though. As with summary, a paraphrase should be written in the reader's own words, while still retaining the meaning of the original source. The main difference between summarizing and paraphrasing is that a summary would be appropriate for a much larger text, while paraphrase might focus on just a few lines of text. Effective paraphrasing will indicate an understanding of the original source, yet still help the reader expand on their interpretation. A paraphrase should neither add new information nor remove essential facts that change the meaning of the source.

Understanding Relationships

In order to better comprehend more complex texts, readers strive to draw connections between ideas or events. Authors often have a main idea or argument that is supported by ideas, facts, or expert opinion. These relationships that are built into writing can take on several different forms.

Depending on the main argument of an informational text, authors may choose to employ a variety of relationship techniques. But before relationships can be developed in writing, the author needs to get organized. What is the main idea or argument? How does the author plan to support that idea? Once the author has a clear picture of what they would like to focus on, they need to build transitions from one idea to the next. Learning the importance of transitioning from one sentence to another, from one paragraph to another, and from one idea to another will not only strengthen the validity of the writing but will also enhance the reader's comprehension.

When transitioning from one sentence to another, authors employ specific connecting words that emphasize the relationships between sentences. Taking the time to consider these transitional words can make the difference between a choppy or confusing paragraph and a well-written one. Consider the following:

> When I was growing up, the neighborhood had kids at every turn. I lived in a townhouse then. In my adult years, I live in a quiet suburb with hardly any children.

> When I was growing up, I lived in a neighborhood where there were kids at every turn. In contrast to my younger years, my adult years are unfolding in a quiet suburb with hardly any children.

Notice how the first example, although written coherently, employs sentences that are somehow disjointed. The transition between the statements is far from smooth. However, in the second example, the simple addition of the phrase "in contrast" connects the two parts of the writer's life and allows the reader to fully comprehend the text.

Learning to transition from sentence to sentence, from paragraph to paragraph, and throughout any piece of writing is an essential skill that helps authors to demonstrate similarities, differences, and relationships, and it helps readers to strengthen comprehension.

Drawing Logical Inferences and Conclusions

Making an **inference** from a selection means to make an educated guess from the passage read. Inferences should be conclusions based off of sound evidence and reasoning. When multiple-choice test questions ask about the logical conclusion that can be drawn from reading text, the test taker must identify which choice will unavoidably lead to that conclusion. In order to eliminate the incorrect choices, the test taker should come up with a hypothetical situation wherein an answer choice is true, but the conclusion is not true. For example, here is an example with three answer choices:

> Fred purchased the newest PC available on the market. Therefore, he purchased the most expensive PC in the computer store.

> What can one assume for this conclusion to follow logically?

> a. Fred enjoys purchasing expensive items.
> b. PCs are some of the most expensive personal technology products available.
> c. The newest PC is the most expensive one.

The premise of the text is the first sentence: Fred purchased the newest PC. The conclusion is the second sentence: Fred purchased the most expensive PC. Recent release and price are two different factors; the difference between them is the logical gap. To eliminate the gap, one must equate whatever new information the conclusion introduces with the pertinent information the premise has stated. This example simplifies the process by having only one of each: one must equate product recency with product price. Therefore, a possible bridge to the logical gap could be a sentence stating that the newest PCs always cost the most.

Making Predictions and Inferences

One technique authors often use to make their fictional stories more interesting is not giving away too much information by providing hints and description. It is then up to the reader to draw a conclusion about the author's meaning by connecting textual clues with the reader's own pre-existing experiences and knowledge. Drawing conclusions is an important reading strategy for understanding what is occurring in a text. Rather than directly stating who, what, where, when, or why, authors often describe story elements. Then, readers must draw conclusions to understand significant story components. As they go through a text, readers can think about the setting, characters, plot, problem, and solution; whether the author provided any clues for consideration; and combine any story clues with their existing knowledge and experiences to draw conclusions about what occurs in the text.

Making Predictions

Before and during reading, readers can apply the reading strategy of making predictions about what they think may happen next. For example, what plot and character developments will occur in fiction? What points will the author discuss in nonfiction? Making predictions about portions of text they have not yet read prepares readers mentally for reading, and also gives them a purpose for reading. To inform and make predictions about text, the reader can do the following:

- Consider the title of the text and what it implies
- Look at the cover of the book
- Look at any illustrations or diagrams for additional visual information
- Analyze the structure of the text
- Apply outside experience and knowledge to the text

Readers may adjust their predictions as they read. Reader predictions may or may not come true in text.

Making Inferences

Authors describe settings, characters, character emotions, and events. Readers must infer to understand text fully. Inferring enables readers to figure out meanings of unfamiliar words, make predictions about upcoming text, draw conclusions, and reflect on reading. Readers can infer about text before, during, and after reading. In everyday life, we use sensory information to infer. Readers can do the same with text. When authors do not answer all reader questions, readers must infer by saying "I think....This could be....This is because....Maybe....This means....I guess..." etc. Looking at illustrations, considering characters' behaviors, and asking questions during reading facilitate inference. Taking clues from text and connecting text to prior knowledge help to draw conclusions. Readers can infer word meanings, settings, reasons for occurrences, character emotions, pronoun referents, author messages, and answers to questions unstated in text. To practice inference, students can read sentences written/selected by the instructor, discuss the setting and character, draw conclusions, and make predictions.

Making inferences and drawing conclusions involve skills that are quite similar: both require readers to fill in information the author has omitted. Authors may omit information as a technique for inducing readers to discover the outcomes themselves; or they may consider certain information unimportant; or they may assume their reading audience already knows certain information. To make an inference or draw a conclusion about text, readers should observe all facts and arguments the author has presented and consider what they already know from their own personal experiences. Reading students taking multiple-choice tests that refer to text passages can determine correct and incorrect choices based on the information in the passage. For example, from a text passage describing an individual's signs of anxiety while unloading groceries and nervously clutching their wallet at a grocery store checkout, readers can infer or conclude that the individual may not have enough money to pay for everything.

Understanding Sequential, Comparative, and Cause-Effect Relationships

Recognizing Events in a Sequence

Sequence structure is the order of events in which a story or information is presented to the audience. Sometimes the text will be presented in chronological order, or sometimes it will be presented by displaying the most recent information first, then moving backwards in time. The sequence structure depends on the author, the context, and the audience. The structure of a text also depends on the genre in which the text is written. Is it literary fiction? Is it a magazine article? Is it instructions for how to complete a certain task? Different genres will have different purposes for switching up the sequence of their writing.

Narrative Structure

The structure presented in literary fiction is also known as **narrative structure**. Narrative structure is the foundation on which the text moves. The basic ways for moving the text along are in the plot and the setting. The plot is the sequence of events in the narrative that move the text forward through cause and effect. The setting of a story is the place or time period in which the story takes place. Narrative structure has two main categories: linear and nonlinear.

Linear narrative is a narrative told in chronological order. Traditional linear narratives will follow the plot diagram below depicting the narrative arc. The narrative arc consists of the exposition, conflict, rising action, climax, falling action, and resolution.

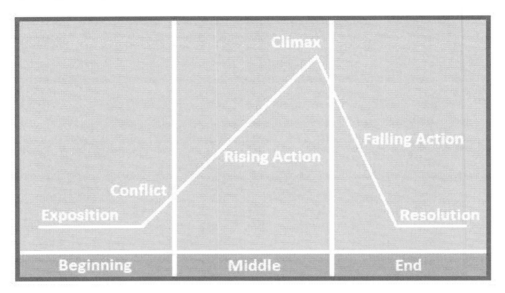

- Exposition: The exposition is in the beginning of a narrative and introduces the characters, setting, and background information of the story. The importance of the exposition lies in its framing of the upcoming narrative. Exposition literally means "a showing forth" in Latin.

- Conflict: The conflict, in a traditional narrative, is presented toward the beginning of the story after the audience becomes familiar with the characters and setting. The conflict is a single instance between characters, nature, or the self, in which the central character is forced to make a decision or move forward with some kind of action. The conflict presents something for the main character, or protagonist, to overcome.

- Rising Action: The rising action is the part of the story that leads into the climax. The rising action will feature the development of characters and plot, and creates the tension and suspense that eventually lead to the climax.

- Climax: The climax is the part of the story where the tension produced in the rising action comes to a culmination. The climax is the peak of the story. In a traditional structure, everything before the climax builds up to it, and everything after the climax falls from it. It is the height of the narrative, and is usually either the most exciting part of the story or is marked by some turning point in the character's journey.

- Falling Action: The falling action happens as a result of the climax. Characters continue to develop, although there is a wrapping up of loose ends here. The falling action leads to the resolution.

- Resolution: The resolution is where the story comes to an end and usually leaves the reader with the satisfaction of knowing what happened within the story and why. However, stories do not always end in this fashion. Sometimes readers can be confused or frustrated at the end from lack of information or the absence of a happy ending.

A **nonlinear narrative** deviates from the traditional narrative in that it does not always follow the traditional plot structure of the narrative arc. Nonlinear narratives may include structures that are disjointed, circular, or disruptive, in the sense that they do not follow chronological order, but rather a nontraditional order of structure. *In medias res* is an example of a structure that predates the linear narrative. *In medias res* is Latin for "in the middle of things," which is how many ancient texts, especially epic poems, began their story, such as Homer's *Iliad*. Instead of having a clear exposition with a full development of characters, they would begin right in the middle of the action.

Modernist texts in the late nineteenth and early twentieth century are known for their experimentation with disjointed narratives, moving away from traditional linear narrative. Disjointed narratives are depicted in novels like *Catch 22*, where the author, Joseph Heller, structures the narrative based on free association of ideas rather than chronology. Another nonlinear narrative can be seen in the novel *Wuthering Heights*, written by Emily Bronte, which disrupts the chronological order by being told retrospectively after the first chapter. There seem to be two narratives in *Wuthering Heights* working at the same time: a present narrative as well as a past narrative. Authors employ disrupting narratives for various reasons; some use it for the purpose of creating situational irony for the readers, while some use it to create a certain effect in the reader, such as excitement, or even a feeling of discomfort or fear.

Sequence Structure in Technical Documents
The purpose of technical documents, such as instructions manuals, cookbooks, or "user-friendly" documents, is to provide information to users as clearly and efficiently as possible. In order to do this, the sequence structure in technical documents that should be used is one that is as straightforward as possible. This usually involves some kind of chronological order or a direct sequence of events. For example, someone who is reading an instruction manual on how to set up their Smart TV wants directions in a clear, simple, straightforward manner that does not leave the reader to guess at the proper sequence or lead to confusion.

Sequence Structure in Informational Texts
The structure in informational texts depends again on the genre. For example, a newspaper article may start by stating an exciting event that happened, and then move on to talk about that event in chronological order. Many informational texts also use **cause and effect structure**, which describes an event and then identifies reasons for why that event occurred. Some essays may write about their subjects by way of **comparison and contrast**, which is a structure that compares two things or contrasts them to highlight their differences. Other documents, such as proposals, will have a **problem to solution structure**, where the document highlights some kind of problem and then offers a solution toward the end. Finally, some informational texts are written with lush details and description in order to captivate the audience, allowing them to visualize the information presented to them. This type of structure is known as **descriptive**.

Comparative and Cause-and-Effect Relationships
Authors employ the compare and contrast strategy to point out the differences or similarities between two subjects. Readers must learn to use comparative thinking to comprehend and evaluate these similarities and differences, but readers have been using comparative thinking throughout their lives. Humans naturally compare and contrast in their everyday lives. Do we relate better to our mothers or our fathers? Should we attend the local college or the out-of-state option? Is it better to own a cat or a dog?

Throughout our lives, people compare and contrast, and in writing, it is no different. Consider the following sentence:

> Although it has been common practice in the United States to encourage the consumption of meat in order to ensure adequate protein levels, more and more people are turning to a plant-based diet, which, they argue, contains healthy levels of protein.

Clearly, the author is comparing and contrasting the two ways in which humans can absorb protein—from meat or from plant-based foods. When students read material that is being compared and contrasted, they begin to use higher-level thinking. The compare and contrast strategy helps readers analyze pairs of ideas, opinions, arguments, or facts, and it improves comprehension by emphasizing important details. Comparing and contrasting can make abstract ideas more concrete, and comparing and contrasting can even strengthen a student's writing skills. Comparing and contrasting helps readers organize information and develop their ideas with greater clarity.

Another common writing strategy involves cause and effect. Cause and effect may be the most recognizable of these relationships and is often used in informational texts. The "cause" refers to the reason why something happened, and the "effect" refers to what happened as a result of the cause. Consider the following sentences:

> I was late for work.

> My alarm was not set.

There is clearly a relationship between these two sentences. The cause of the individual's having been late for work was the unset alarm. The effect the unset alarm had was the individual's arriving late for work. In order to connect these two sentences into a cause-and-effect sentence, an author could write:

> Since my alarm was not set, I was late for work.

Some of the most commonly used keywords that help to identify cause-and-effect relationships in writing include, *because, since, so, if, then, before*, and, *after*. But authors do not always make the relationship this simple to detect. Oftentimes, the cause follows the effect. Consider the following sentence:

> The college tuition significantly decreased after the government's announcement of improved funding.

The cause in this sentence is the government's announcement of improved funding. The effect of this improved funding is that college tuition has significantly decreased.

Understanding what happened and why helps to strengthen reading comprehension, develop skills that identify patterns in writing, and enhances the ability to explain and analyze the writing.

Craft and Structure

Determining Word and Phrase Meanings

When readers encounter an unfamiliar word in text, they can use the surrounding context—the overall subject matter, specific chapter/section topic, and especially the immediate sentence context. Among others, one category of context clues is grammar. For example, the position of a word in a sentence and its relationship to the other words can help the reader establish whether the unfamiliar word is a verb, a noun, an adjective, an adverb, etc. This narrows down the possible meanings of the word to one part of speech. However, this may be insufficient. In a sentence that many birds *migrate* twice yearly, the reader can determine the word is a verb, and probably does not mean eat or drink; but it could mean travel, mate, lay eggs, hatch, molt, etc.

Some words can have a number of different meanings depending on how they are used. For example, the word *fly* has a different meaning in each of the following sentences:

- "His trousers have a fly on them."
- "He swatted the fly on his trousers."
- "Those are some fly trousers."
- "They went fly fishing."
- "She hates to fly."
- "If humans were meant to fly, they would have wings."

As strategies, readers can try substituting a familiar word for an unfamiliar one and see whether it makes sense in the sentence. They can also identify other words in a sentence, offering clues to an unfamiliar word's meaning.

Readers can often figure out what unfamiliar words mean without interrupting their reading to look them up in dictionaries by examining context. **Context** includes the other words or sentences in a passage. One common context clue is the root word and any affixes (prefixes/suffixes). Another common context clue is a synonym or definition included in the sentence. Sometimes both exist in the same sentence. Here's an example:

Scientists who study birds are *ornithologists*.

Many readers may not know the word *ornithologist*. However, the example contains a definition (scientists who study birds). The reader may also have the ability to analyze the suffix (*-logy*, meaning the study of) and root (*ornitho-*, meaning bird).

Another common context clue is a sentence that shows differences. Here's an example:

Birds *incubate* their eggs outside of their bodies, unlike mammals.

Some readers may be unfamiliar with the word *incubate*. However, since we know that "unlike mammals," birds incubate their eggs outside of their bodies, we can infer that *incubate* has something to do with keeping eggs warm outside the body until they are hatched.

In addition to analyzing the etymology of a word's root and affixes and extrapolating word meaning from sentences that contrast an unknown word with an antonym, readers can also determine word meanings from sentence context clues based on logic.

Here's an example:

Birds are always looking out for predators that could attack their young.

The reader who is unfamiliar with the word *predator* could determine from the context of the sentence that predators usually prey upon baby birds and possibly other young animals. Readers might also use the context clue of etymology here, as *predator* and *prey* have the same root.

Denotation and Connotation

Denotation refers to a word's explicit definition, like that found in the dictionary. Denotation is often set in comparison to connotation. **Connotation** is the emotional, cultural, social, or personal implication associated with a word. Denotation is more of an objective definition, whereas connotation can be more subjective, although many connotative meanings of words are similar for certain cultures. The denotative meanings of words are usually based on facts, and the connotative meanings of words are usually based on emotion. Here are some examples of words and their denotative and connotative meanings in Western culture:

Word	Denotative Meaning	Connotative Meaning
Home	A permanent place where one lives, usually as a member of a family.	A place of warmth; a place of familiarity; comforting; a place of safety and security. "Home" usually has a positive connotation.
Snake	A long reptile with no limbs and strong jaws that moves along the ground; some snakes have a poisonous bite.	An evil omen; a slithery creature (human or nonhuman) that is deceitful or unwelcome. "Snake" usually has a negative connotation.
Winter	A season of the year that is the coldest, usually from December to February in the northern hemisphere and from June to August in the southern hemisphere.	Circle of life, especially that of death and dying; cold or icy; dark and gloomy; hibernation, sleep, or rest. Winter can have a negative connotation, although many who have access to heat may enjoy the snowy season from their homes.

Analyzing an Author's Word Choice

An author's word choice can also affect the style, tone, and mood of the text. Word choices, grammatical choices, and syntactical choices can help the audience figure out the scope, purpose, and emphasis. These choices—embedded in the words and sentences of the passage (i.e., the "parts")—help paint the intentions and goals of the author (i.e., the "whole"). For instance, if an author is using strong language like *enrage, ignite, infuriate,* and *antagonize,* then they may be cueing the reader into their own rage or they may be trying to incite anger in other. Likewise, if an author is continually using rapid, simple sentences, he or she might be trying to incite excitement and nervousness. These different choices and styles affect the overall message, or purpose. Sometimes the subject matter or audience will be discussed explicitly, but often readers have to decode the passage, or "break it down," to find the target audience and intentions. Meanwhile, the impact of the article can be personal or historical, for example, depending upon the passage—it can either "speak" to you personally or "capture" an historical era.

Analyzing Text Structure

Text structure is the way in which the author organizes and presents textual information so readers can follow and comprehend it. One kind of text structure is sequence. This means the author arranges the text in a logical order from beginning to middle to end. There are three types of sequences:

- Chronological: ordering events in time from earliest to latest

- Spatial: describing objects, people, or spaces according to their relationships to one another in space

- Order of Importance: addressing topics, characters, or ideas according to how important they are, from either least important to most important

Chronological sequence is the most common sequential text structure. Readers can identify sequential structure by looking for words that signal it, like *first, earlier, meanwhile, next, then, later, finally;* and specific times and dates the author includes as chronological references.

Problem-Solution Text Structure

The problem-solution text structure organizes textual information by presenting readers with a problem and then developing its solution throughout the course of the text. The author may present a variety of alternatives as possible solutions, eliminating each as they are found unsuccessful, or gradually leading up to the ultimate solution. For example, in fiction, an author might write a murder mystery novel and have the character(s) solve it through investigating various clues or character alibis until the killer is identified. In nonfiction, an author writing an essay or book on a real-world problem might discuss various alternatives and explain their disadvantages or why they would not work before identifying the best solution. For scientific research, an author reporting and discussing scientific experiment results would explain why various alternatives failed or succeeded.

Comparison-Contrast Text Structure

Comparison identifies similarities between two or more things. **Contrast** identifies differences between two or more things. Authors typically employ both to illustrate relationships between things by highlighting their commonalities and deviations. For example, a writer might compare Windows and Linux as operating systems, and contrast Linux as free and open-source vs. Windows as proprietary. When writing an essay, sometimes it is useful to create an image of the two objects or events you are comparing or contrasting. Venn diagrams are useful because they show the differences as well as the similarities between two things. Once you've seen the similarities and differences on paper, it might be helpful to create an outline of the essay with both comparison and contrast. Every outline will look different, because every two or more things will have a different number of comparisons and contrasts. Say you are trying to compare and contrast carrots with sweet potatoes. Here is an example of a compare/contrast outline using those topics:

- Introduction: Talk about why you are comparing and contrasting carrots and sweet potatoes. Give the thesis statement.

- Body paragraph 1: Sweet potatoes and carrots are both root vegetables (similarity)

- Body paragraph 2: Sweet potatoes and carrots are both orange (similarity)

- Body paragraph 3: Sweet potatoes and carrots have different nutritional components (difference)

- Conclusion: Restate the purpose of your comparison/contrast essay.

Of course, if there is only one similarity between your topics and two differences, you will want to rearrange your outline. Always tailor your essay to what works best with your topic.

Descriptive Text Structure

Description can be both a type of text structure and a type of text. Some texts are descriptive throughout entire books. For example, a book may describe the geography of a certain country, state, or region, or tell readers all about dolphins by describing many of their characteristics. Many other texts are not descriptive throughout, but use descriptive passages within the overall text. The following are a few examples of descriptive text:

- When the author describes a character in a novel
- When the author sets the scene for an event by describing the setting
- When a biographer describes the personality and behaviors of a real-life individual
- When a historian describes the details of a particular battle within a book about a specific war
- When a travel writer describes the climate, people, foods, and/or customs of a certain place

A hallmark of description is using sensory details, painting a vivid picture so readers can imagine it almost as if they were experiencing it personally.

Cause and Effect Text Structure
When using cause and effect to extrapolate meaning from text, readers must determine the cause when the author only communicates effects. For example, if a description of a child eating an ice cream cone includes details like beads of sweat forming on the child's face and the ice cream dripping down her hand faster than she can lick it off, the reader can infer or conclude it must be hot outside. A useful technique for making such decisions is wording them in "If...then" form, e.g. "If the child is perspiring and the ice cream melting, it may be a hot day." Cause and effect text structures explain why certain events or actions resulted in particular outcomes. For example, an author might describe America's historical large flocks of dodo birds, the fact that gunshots did not startle/frighten dodos, and that because dodos did not flee, settlers killed whole flocks in one hunting session, explaining how the dodo was hunted into extinction.

Understanding Authorial Purpose and Perspective

Authors may have many purposes for writing a specific text. Their purposes may be to try and convince readers to agree with their position on a subject, to impart information, or to entertain. Other writers are motivated to write from a desire to express their own feelings. Authors' purposes are their reasons for writing something. A single author may have one overriding purpose for writing or multiple reasons. An author may explicitly state their intention in the text, or the reader may need to infer that intention. Those who read reflectively benefit from identifying the purpose because it enables them to analyze information in the text. By knowing why the author wrote the text, readers can glean ideas for how to approach it. The following is a list of questions readers can ask in order to discern an author's purpose for writing a text:

- From the title of the text, why do you think the author wrote it?
- Was the purpose of the text to give information to readers?
- Did the author want to describe an event, issue, or individual?
- Was it written to express emotions and thoughts?
- Did the author want to convince readers to consider a particular issue?
- Was the author primarily motivated to write the text to entertain?
- Why do you think the author wrote this text from a certain point of view?
- What is your response to the text as a reader?
- Did the author state their purpose for writing it?

Readers should read to interpret information rather than simply content themselves with roles as text consumers. Being able to identify an author's purpose efficiently improves reading comprehension, develops critical thinking, and makes readers more likely to consider issues in depth before accepting writer viewpoints. Authors of fiction frequently write to entertain readers. Another purpose for writing fiction is making a political statement; for example, Jonathan Swift wrote "A Modest Proposal" (1729) as a political satire. Another purpose for writing fiction as well as nonfiction is to persuade readers to take some action or further a particular cause. Fiction authors and poets both frequently write to evoke certain moods; for example, Edgar Allan Poe wrote novels, short stories, and poems that evoke moods of gloom, guilt, terror, and dread. Another purpose of poets is evoking certain emotions: love is popular, as in Shakespeare's sonnets and numerous others. In "The Waste Land" (1922), T.S. Eliot evokes society's alienation, disaffection, sterility, and fragmentation.

Authors seldom directly state their purposes in texts. Some readers may be confronted with nonfiction texts such as biographies, histories, magazine and newspaper articles, and instruction manuals, among others. To identify the purpose in nonfiction texts, students can ask the following questions:

- Is the author trying to teach something?
- Is the author trying to persuade the reader?
- Is the author imparting factual information only?
- Is this a reliable source?
- Does the author have some kind of hidden agenda?

To apply author purpose in nonfictional passages, readers can also analyze sentence structure, word choice, and transitions to answer the aforementioned questions and to make inferences. For example, authors wanting to convince readers to view a topic negatively often choose words with negative connotations.

Narrative Writing

Narrative writing tells a story. The most prominent examples of narrative writing are fictional novels. Here are some examples:

- Mark Twain's The Adventures of Tom Sawyer and The Adventures of Huckleberry Finn
- Victor Hugo's *Les Misérables*
- Charles Dickens' Great Expectations, David Copperfield, and A Tale of Two Cities
- Jane Austen's Northanger Abbey, Mansfield Park, Pride and Prejudice, and Sense and Sensibility
- Toni Morrison's Beloved, The Bluest Eye, and Song of Solomon
- Gabriel García Márquez's One Hundred Years of Solitude and Love in the Time of Cholera

Some nonfiction works are also written in narrative form. For example, some authors choose a narrative style to convey factual information about a topic, such as a specific animal, country, geographic region, and scientific or natural phenomenon.

Since narrative is the type of writing that tells a story, it must be told by someone, who is the narrator. The narrator may be a fictional character telling the story from their own viewpoint. This narrator uses the first person (*I, me, my, mine* and *we, us, our,* and *ours*). The narrator may simply be the author; for example, when Louisa May Alcott writes "Dear reader" in *Little Women*, she (the author) addresses us as readers. In this case, the novel is typically told in third person, referring to the characters as he, she, they, or them. Another more common technique is the omniscient narrator; i.e. the story is told by an unidentified individual who sees and knows everything about the events and characters—not only their externalized actions, but also their internalized feelings and thoughts. Second person, i.e. writing the story by addressing readers as "you" throughout, is less frequently used.

Expository Writing

Expository writing is also known as informational writing. Its purpose is not to tell a story as in narrative writing, to paint a picture as in descriptive writing, or to persuade readers to agree with something as in argumentative writing. Rather, its point is to communicate information to the reader. As such, the point of view of the author will necessarily be more objective. Whereas other types of writing appeal to the reader's emotions, appeal to the reader's reason by using logic, or use subjective descriptions to sway the reader's opinion or thinking, expository writing seeks to do none of these but simply to provide facts, evidence, observations, and objective descriptions of the subject matter. Some examples of expository writing include research reports, journal articles, articles and books about historical events or periods, academic subject textbooks, news articles and other factual journalistic reports, essays, how-to articles, and user instruction manuals.

Technical Writing

Technical writing is similar to expository writing in that it is factual, objective, and intended to provide information to the reader. Indeed, it may even be considered a subcategory of expository writing. However, technical writing differs from expository writing in that (1) it is specific to a particular field, discipline, or subject; and (2) it uses the specific technical terminology that belongs only to that area. Writing that uses technical terms is intended only for an audience familiar with those terms. A primary example of technical writing today is writing related to computer programming and use.

Persuasive Writing

Persuasive writing is intended to persuade the reader to agree with the author's position. It is also known as argumentative writing. Some writers may be responding to other writers' arguments, in which case they make reference to those authors or text and then disagree with them. However, another common technique is for the author to anticipate opposing viewpoints in general, both from other authors and from the author's own readers. The author brings up these opposing viewpoints, and then refutes them before they can even be raised, strengthening the author's argument. Writers persuade readers by appealing to their reason, which Aristotle called *logos;* appealing to emotion, which Aristotle called *pathos;* or appealing to readers based on the author's character and credibility, which Aristotle called *ethos.*

Analyzing Characters' Points of View

In fiction, authors either write from the first-, second-, or third-person point of view. Throughout a literary work, authors may choose to write exclusively from one point of view, two points of view, or even all three. Analyzing points of view often leads readers to reflect and consider various perspectives on a given subject.

First Person

First-person singular point of view becomes apparent when the author uses the pronouns "I," "me," "my," and "mine." The use of pronouns "we," "us," "our," and "ours" indicates the use of the first-person plural. Authors often choose first-person point of view to develop a close connection with the audience. First-person point of view brings a familiar and human feel to the writing, to which many readers relate. Often filled with subjective messaging, first-person point of view strives to connect with the readers on a personal level. Consider the following first-person point of view:

> "It was Sunday, the best day of the week. After church, Mama would take me straight to Grandma's house for cookies and tea. We would rock on the rocking chair on the front porch as if we didn't have a care in the world—truth be told, we really didn't have a care in the world."

This example demonstrates a clear example of how an author strives to pull in the reader with the use of first-person point of view. Readers connect to the sentimentality and develop a sense of nostalgia.

Second Person

Second-person point of view employs the pronouns "you," "your," and "yours." Only occasionally used in fiction, second-person point of view requires a lot more effort to develop. When authors want to fully immerse their audience in the experiences unfolding in the story, or when they wish for the audience to imagine themselves in that exact place and time, feeling that exact way, they may choose a second-person point of view. Consider the following passage:

> "Imagine you were at the site when the first thunderbolt fell from the sky. You look up and cannot believe your eyes. At first, you are mesmerized, but that feeling quickly morphs into shock. You look to your left, then to your right, because in that moment, you do not want to be alone. You want nothing more than to share these sensations with someone you love."

Second-person point of view, however, is difficult to sustain for a long period of time, especially in fiction, and it is better used for only brief moments when authors wish to plunge their audience directly into the storyline.

Third Person

When written from the third-person point of view, the writing can sometimes feel distant. The reader is somewhat removed from the experiences taking place. In literature, third-person points of view are developed with the use of a narrator who acts as a person on the outside looking in and giving play-by-play accounts of what is taking place. Pronouns "he," "she," "they," and even "it" can be used to describe the scenes in the story. Consider the following passage:

> "Emily knew it was just a matter of time before she would have to leave. She heard the clock ticking in the big, empty hallway, and it seemed as loud as a thousand church bells. She sat—completely still—until the clock struck twelve. Then she drew a deep breath, stood, picked up her bags, and left. 'Soon, it will all be over', she whispered to herself."

Although the narrator is describing Emily from a distance, and readers also feel somewhat removed, they can still feel that sense of fear, or perhaps anxiety, as Emily awaits the moment when she must leave.

Analyzing point of view is an essential skill for readers to develop in order to gain a deeper understanding of the storyline, as well as the different characters who all play a role in the story's development.

Interpreting Authorial Decisions Rhetorically

One of the freedoms in reading is to derive a unique perspective on the author's intent. Writing is an art form and can have many different interpretations. Often, readers who read the same literary work may have varying opinions regarding the author's intent. In fact, their varying interpretations might also differ from the author's intended message. Reading literature is less about being right than about striving to derive meaning from the message. The beauty of literature, as in any art form, is that it is open to interpretation.

There have been many theories about the intended meaning of Lewis Carroll's *Alice's Adventures in Wonderland*, from a bizarre take on the world brought on by drug use, to an obsession with food and drink—and many other interpretations as well. However, the author himself said that the intent was nothing more than to entertain a child friend by creating a dreamlike, fantastical tale. Does that mean, then, that readers should stop imagining, should stop analyzing, and should simply accept the author's intended meaning? Although it should be respected, the author's intended meaning may not be the only interpretation. Literature can often take on a life of its own, and each reader is free to interpret literature in their own unique way, while keeping an open mind on other perspectives, particularly that of the author. Authors may learn a great deal about their work of art through the eyes of their readers, and authors may develop different perspectives based on their readers' keen observations and discoveries. True artists appreciate different interpretations of their work of art, and they regard the varied interpretations as something to be celebrated. Readers and authors affect one another, learn from one another, and become more skilled in their art when they allow their own interpretations to be analyzed.

Differentiating Between Various Perspectives

"Point of view" refers to the type of narration the author employs in a given story. "Perspective" refers to how characters perceive what is happening within the story. The characters' perspectives reveal their attitudes and help to shape their unique personalities. Consider the following scenario:

> "The family grabbed their snacks and blankets, loaded up the van, and headed out to the neighborhood park, even though Suki would have preferred to stay home. Once they settled in at their spot on the grass, the celebration was about to start. Within minutes, the fireworks began—crack, bang, pop! Hendrix jumped up and down with glee, Suki angrily put down her phone, and the dog yelped and buried its head under the blankets."

Each character was experiencing the same event—fireworks—and yet each character had a different reaction. Hendrix seems excited, Suki, angry, and the dog, frightened. No *one* perspective is the "right" perspective, just as no particular perspective is wrong; they are simply perspectives. What makes characters unique within a story are their unique perspectives. When authors develop characters with unique personalities and differing perspectives, stories are not only more believable, but they are more alive, more colorful, and more interesting. If all characters had the same one-dimensional perspective, the story would likely be quite dull. There would never be a protagonist or antagonist, and there would be no reason to examine why each character acts and reacts to situations in such unique ways. Differentiating between various perspectives in a story can also lead to a much deeper understanding. For instance, it seems relatively easy to consider the perspective of the protagonist in any story since most readers connect with good and reject evil. But readers might wish to explore the story through the eyes of the antagonist. They might want to discover how the antagonist ended up so villainous, what events led to their corruption, and what, if anything, might lead them back to truth and justice.

Perspectives are how individuals see the world in which they live, and they are often formed from the individual's unique life experiences, their morals, and their values. Differentiating between various perspectives in literature helps readers to develop a greater appreciation for the story and for each character that helps to shape that story.

Comparing Different Sources of Information

Identifying Specific Information from a Printed Communication
Business Memos
Whereas everyday office memos were traditionally typed on paper, photocopied, and distributed, today they are more often typed on computers and distributed via e-mail, both interoffice and externally. Technology has thus made these communications more immediate. It is also helpful for people to read carefully and be familiar with memo components. For example, e-mails automatically provide the same "To:, From:, and Re:" lines traditionally required in paper memos, and in corresponding places—the top of the page/screen. Readers should observe whether "To: names/positions" include all intended recipients in case of misdirection errors or omitted recipients. "From:" informs sender level, role, and who will receive

responses when people click "Reply." Users must be careful not to click "Reply All" unintentionally. They should also observe the "CC:" line, typically below "Re:," showing additional recipients.

Classified Ads
Classified advertisements include "Help Wanted" ads informing readers of positions open for hiring, real estate listings, cars for sale, and home and business services available. Traditional ads in newspapers had to save space, and this necessity has largely transferred to online ads. Because of needing to save space, advertisers employ many abbreviations. For example, here are some examples of abbreviations:

- FT=full-time
- PT=part-time
- A/P=accounts payable
- A/R=accounts receivable
- Asst.=assistant
- Bkkg.=bookkeeping
- Comm.=commission
- Bet.=between
- EOE=equal opportunity employer
- G/L=general ledger
- Immed.=immediately
- Exc.=excellent
- Exp.=experience
- Eves.=evenings
- Secy.=secretary
- Temp=temporary
- Sal=salary
- Req=required
- Refs=references
- Wk=week or work
- WPM=words per minute

Classified ads frequently use abbreviations to take up less space, both on paper and digitally on websites. Those who read these ads will find it less confusing if they learn some common abbreviations used by businesses when advertising job positions. Here are some examples:

- Mgt.=management
- Mgr.=manager
- Mfg.=manufacturing
- Nat'l=national
- Dept.=department
- Min.=minimum
- Yrs.=years
- Nec=necessary
- Neg=negotiable
- Oppty=opportunity
- O/T=overtime
- K=1,000

Readers of classified ads may focus on certain features to the exclusion of others. For example, if a reader sees the job title or salary they are seeking, or notices the experience, education, degree, or other credentials required match their own qualifications perfectly, they may fail to notice other important information, like "No benefits." This is important because the employers are disclosing that they will not provide health insurance, retirement accounts, paid sick leave, paid maternity/paternity leave, paid vacation, etc. to any employee whom they hire. Someone expecting a traditional 9 to 5 job who fails to observe that an ad states "Evenings" or just the abbreviation "Eves" will be disappointed, as will the applicant who overlooks a line saying "Some evenings and weekends reqd." Applicants overlooking information like "Apply in person" may e-mail or mail their resumes and receive no response. The job hopeful with no previous experience and one reference must attend to information like "Minimum 5 yrs. exp, 3 refs," meaning they likely will not qualify.

Employment Ads
Job applicants should pay attention to the information included in classified employment ads. On one hand, they do need to believe and accept certain statements, such as "Please, no phone calls," which is frequently used by employers posting ads on Craigslist and similar websites. New applicants just graduated from or still in college will be glad to see "No exp necessary" in some ads, indicating they need no previous work experience in that job category. "FT/PT" means the employer offers the options of working full-time or part-time, another plus for students. On the other hand, ad readers should also take into consideration the fact that many employers list all the attributes of their *ideal* employee, but they do not necessarily expect to find such a candidate. If a potential applicant's education, training, credentials, and experience are not exactly the same as what the employer lists as desired but are not radically different either, it can be productive to apply anyway, while honestly representing one's actual qualifications.

Atlases
A road atlas is a publication designed to assist travelers who are driving on road trips rather than taking airplanes, trains, ships, etc. Travelers use road atlases to determine which routes to take and places to stop; how to navigate specific cities, locate landmarks, estimate mileages and travel times; see photographs of places they plan to visit; and find other travel-related information. One familiar, reputable road atlas is published by the National Geographic Society. It includes detailed, accurate maps of the United States, Canada, and Mexico; historic sites, scenic routes, recreation information, and points of interest; and its Adventure Edition spotlights 100 top U.S. adventure destinations and most popular national parks. The best-selling road atlas in the United States, also probably the best-known and most trusted, is published annually by Rand McNally, which has published road atlases for many years. It includes maps, mileage charts, information on tourism and road construction, maps of individual city details, and the editor's favorite road trips (in the 2016 edition) including recommended points of interest en route.

Owners' Manuals
An owner's manual is typically a booklet, but may also be as short as a page or as long as a book, depending on the individual instance. The purpose of an owner's manual is to give the owner instructions, usually step-by-step, for how to use a specific product or a group or range of products. Manuals accompany consumer products as diverse as cars, computers, tablets, smartphones, printers, home appliances, shop machines, and many others. In addition to directions for operating products, they include important warnings of things *not* to do that pose safety or health hazards or can damage the product and void the manufacturer's product warranty, like immersion in water, exposure to high temperatures, operating something for too long, dropping fragile items, or rough handling. Manuals

teach correct operating practices, sequences, precautions, and cautions, averting many costly and/or dangerous mishaps.

Food Labels

When reading the labels on food products, it is often necessary to interpret the nutrition facts and other product information. Without the consumer's being aware and informed of this, much of this information can be very misleading. For example, a popular brand name of corn chips lists the calories, fat, etc. per serving in the nutrition facts printed on the bag, but on closer inspection, it defines a serving size as six chips—far fewer than most people would consume at a time. Serving sizes and the number of servings per container can be unrealistic. For example, a jumbo muffin's wrapper indicates it contains three servings. Not only do most consumers not divide a muffin and eat only part; but it is moreover rather difficult to cut a muffin into equal thirds. A king-sized package of chili cheese-flavored corn chips says it contains 4.5 servings per container. This is not very useful information, since people cannot divide the package into equal servings and are unlikely to eat four servings and then ½ a serving.

Product Packaging

Consumers today cannot take product labels at face value. While many people do not read or even look at the information on packages before eating their contents, those who do must use more consideration and analysis than they might expect to understand it. For example, a well-known brand of strawberry-flavored breakfast toaster pastry displays a picture of four whole strawberries on the wrapper. While this looks appealing, encouraging consumers to infer the product contains wholesome fruit—and perhaps even believe it contains four whole strawberries—reading the ingredients list reveals it contains only 2 percent or less of dried strawberries. A consumer must be detail-oriented (and curious or motivated enough) to read the full ingredients list, which also reveals unhealthy corn syrup and high fructose corn syrup high on the list after enriched flour. Consumers must also educate themselves about euphemistically misleading terms: "enriched" flour has vitamins and minerals added, but it is refined flour without whole grain, bran, or fiber.

While manufacturers generally provide extensive information printed on their package labels, it is typically in very small print—many consumers do not read it—and even consumers who do read all the information must look for small details to discover that the information is often not realistic. For example, a box of brownie mix lists grams of fat, total calories, and calories from fat. However, by paying attention to small details like asterisks next to this information, and finding the additional information referenced by the asterisks, the consumer discovers that these amounts are for only the dry mix—not the added eggs, oil, or milk. Consumers typically do not eat dry cake mixes, and having to determine and add the fat and calories from the additional ingredients is inconvenient. In another example, a box of macaroni and cheese mix has an asterisk by the fat grams indicating these are for the macaroni only without the cheese, butter, or milk required, which contributes 6.4 times more fat.

Ingredients' Lists

Consumers can realize the importance of reading drug labeling through an analogy: What might occur if they were to read only part of the directions on a standardized test? Reading only part of the directions on medications can have similar, even more serious consequences. Prescription drug packages typically contain inserts, which provide extremely extensive, thorough, detailed information, including results of clinical trials and statistics showing patient responses and adverse effects. Some over-the-counter medications include inserts, and some do not. "Active ingredients" are those ingredients making medication effective. "Inactive ingredients" including flavorings, preservatives, stabilizers, and emulsifiers have purposes, but not to treat symptoms. "Uses" indicates which symptoms a medication is meant to treat. "Directions" tell the dosage, frequency, maximum daily amount, and other requirements, like "Take with food," "Do not operate heavy machinery while using," etc. Drug labels also state how to store the product, like at what temperature or away from direct sunlight or humidity.

Many drugs which were previously available only by doctor prescription have recently become available over the counter without a prescription. While enough years of testing may have determined that these substances typically do not cause serious problems, consumers must nevertheless thoroughly read and understand all the information on the labels before taking them. If they do not, they could still suffer serious harm. For example, some individuals have allergies to specific substances. Both prescription and over-the-counter medication products list their ingredients, including warnings about allergies. Allergic reactions can include anaphylactic shock, which can be fatal if not treated immediately. Also, consumers must read and follow dosing directions: taking more than directed can cause harm, and taking less can be ineffective to treat symptoms. Some medication labels warn not to mix them with certain other drugs to avoid harmful drug interactions. Additionally, without reading ingredients, some consumers take multiple products including the same active ingredients, resulting in overdoses.

Identifying Information from a Graphic Representation of Information
Line Graphs
Line graphs are useful for visually representing data that vary continuously over time, like an individual student's test scores. The horizontal or x-axis shows dates/times; the vertical or y-axis shows point values. A dot is plotted on the point where each horizontal date line intersects each vertical number line, and then these dots are connected, forming a line. Line graphs show whether changes in values over time exhibit trends like ascending, descending, flat, or more variable, like going up and down at different times. For example, suppose a student's scores on the same type of reading test were 75% in October, 80% in November, 78% in December, 82% in January, 85% in February, 88% in March, and 90% in April. A line graph of these scores would look like this:

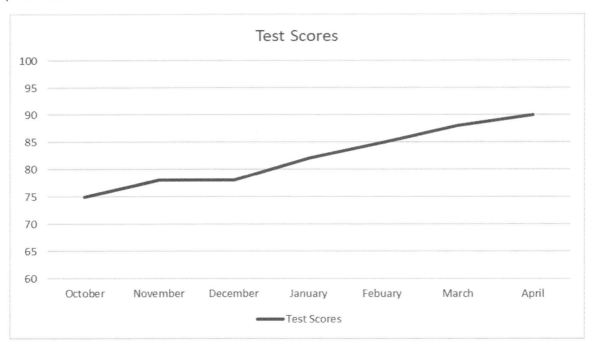

Bar Graphs
Bar graphs feature equally spaced, horizontal or vertical rectangular bars representing numerical values. They can show change over time as line graphs do, but unlike line graphs, bar graphs can also show differences and similarities among values at a single point in time. Bar graphs are also helpful for visually

representing data from different categories, especially when the horizontal axis displays some value that is not numerical, like various countries with inches of annual rainfall. The following is a bar graph that compares different classes and how many books they read:

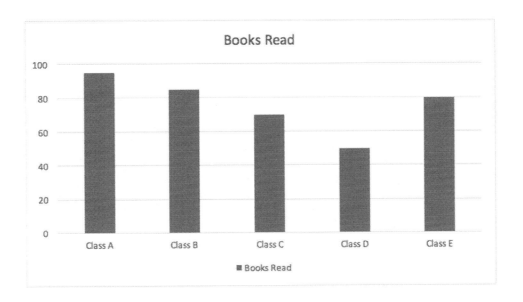

Pie Charts

Pie charts, also called circle graphs, are good for representing percentages or proportions of a whole quantity because they represent the whole as a circle or "pie," with the various proportion values shown as "slices" or wedges of the pie. This gives viewers a clear idea of how much of a total each item occupies. To calculate central angles to make each portion the correct size, multiply each percentage by 3.6 (= 360/100). For example, biologists may have information that 60% of Americans have brown eyes, 20% have hazel eyes, 15% have blue eyes, and 5% have green eyes. A pie chart of these distributions would look like this:

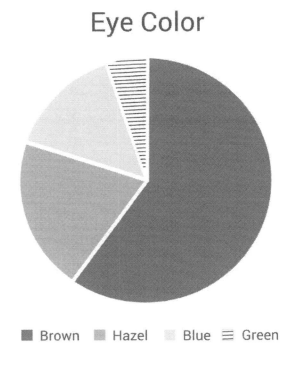

Line Plots

Rather than showing trends or changes over time like line graphs, line plots show the frequency with which a value occurs in a group. Line plots are used for visually representing data sets that total 50 or fewer values. They make visible features like gaps between some data points, clusters of certain numbers/number ranges, and outliers (data points with significantly smaller or larger values than others). For example, the age ranges in a class of nursing students might appear like this in a line plot:

XXXXXXXXXX	XXXXXX	XX	X	XXX	XX	X
18	23	28	33	38	43	48

Pictograms

Magazines, newspapers, and other similar publications designed for consumption by the general public often use pictograms to represent data. Pictograms feature icons or symbols that look like whatever category of data is being counted, like little silhouettes shaped like human beings commonly used to represent people. If the data involve large numbers, like populations, one person symbol might represent one million people, or one thousand, etc. For smaller values, such as how many individuals out of ten fit a given description, one symbol might equal one person. Male and female silhouettes are used to differentiate gender, and child shapes for children. Little clock symbols are used to represent amounts of time, such as a given number of hours; calendar pages might depict months; suns and moons could show days and nights; hourglasses might represent minutes. While pictogram symbols are easily recognizable and appealing to general viewers, one disadvantage is that it is difficult to display partial symbols for in-between quantities.

Integration of Knowledge and Ideas

Understanding Authors' Claims

In academic writing, an author's claim refers to the argument being made, the main idea, or the thesis statement. It is the point that the author chooses to impress upon the audience whether that point is a truth that can be proven or an opinion that can be supported. In order for claims to be valid, they should always be accompanied by supporting evidence. To understand an author's claim, readers first identify the claim and then look for the supporting evidence. Finally, they analyze the evidence to determine its strength. Is the evidence based on popular belief, expert opinion, or empirical data?

But understanding an author's claim isn't just about identifying the main argument. It is also about considering how the author supports that claim and ruling out any bias that would weaken the argument. Is the author trying to share findings, or is the intent more about persuasion? Even if the author's intent is to persuade, the argument can still be valid. After all, a well-supported claim that addiction to drugs and alcohol negatively affects a person's social, emotional, and physical wellbeing might be persuasive, and it is certainly a valid claim. The three modes of persuasion involve:

- Ethos, which refers to the author's credibility or character
- Pathos, which refers to emotion
- Logos, which refers to logic

Examining the mode the author uses results in a greater understanding of the author's claims, enabling the reader to determine the claim's credibility. To further understand the author's claim, it also helps to identify the author's intended audience. Is the author writing directly to scholars or to students? Is the message intended to effect governmental or social change?

Authors present their claim and strive to support it with reasons and evidence. It is the reader's responsibility to identify the claim and the supporting evidence, and to evaluate whether the supporting evidence is strong enough to hold up the author's claim.

Differentiating Between Facts and Opinions

Facts and Opinions
A **fact** is a statement that is true empirically or an event that has actually occurred in reality, and can be proven or supported by evidence; it is generally objective. In contrast, an **opinion** is subjective, representing something that someone believes rather than something that exists in the absolute. People's individual understandings, feelings, and perspectives contribute to variations in opinion. Though facts are typically objective in nature, in some instances, a statement of fact may be both factual and yet also subjective. For example, emotions are individual subjective experiences. If an individual says that they feel happy or sad, the feeling is subjective, but the statement is factual; hence, it is a subjective fact. In

contrast, if one person tells another that the other is feeling happy or sad—whether this is true or not—that is an assumption or an opinion.

Biases
Biases usually occur when someone allows their personal preferences or ideologies to interfere with what should be an objective decision. In personal situations, someone is biased towards someone if they favor them in an unfair way. In academic writing, being biased in your sources means leaving out objective information that would turn the argument one way or the other. The evidence of bias in academic writing makes the text less credible, so be sure to present all viewpoints when writing, not just your own, so to avoid coming off as biased. Being objective when presenting information or dealing with people usually allows the person to gain more credibility.

Stereotypes
Stereotypes are preconceived notions that place a particular rule or characteristics on an entire group of people. Stereotypes are usually offensive to the group they refer to or allies of that group, and often have negative connotations. The reinforcement of stereotypes isn't always obvious. Sometimes stereotypes can be very subtle and are still widely used in order for people to understand categories within the world. For example, saying that women are more emotional and intuitive than men is a stereotype, although this is still an assumption used by many in order to understand the differences between one another.

Using Evidence to Make Connections Between Different Texts That are Related by Topic

Comparing and Contrasting Themes from Print and Other Sources
The **theme** of a piece of text is the central idea the author communicates. Whereas the topic of a passage of text may be concrete in nature, by contrast the theme is always conceptual. For example, while the topic of Mark Twain's novel *The Adventures of Huckleberry Finn* might be described as something like the coming-of-age experiences of a poor, illiterate, functionally orphaned boy around and on the Mississippi River in 19th-century Missouri, one theme of the book might be that human beings are corrupted by society. Another might be that slavery and "civilized" society itself are hypocritical. Whereas the main idea in a text is the most important single point that the author wants to make, the theme is the concept or view around which the author centers the text.

Throughout time, humans have told stories with similar themes. Some themes are universal across time, space, and culture. These include themes of the individual as a hero, conflicts of the individual against nature, the individual against society, change vs. tradition, the circle of life, coming-of-age, and the complexities of love. Themes involving war and peace have featured prominently in diverse works, like Homer's *Iliad*, Tolstoy's *War and Peace* (1869), Stephen Crane's *The Red Badge of Courage* (1895), Hemingway's *A Farewell to Arms* (1929), and Margaret Mitchell's *Gone with the Wind* (1936). Another universal literary theme is that of the quest. These appear in folklore from countries and cultures worldwide, including the Gilgamesh Epic, Arthurian legend's Holy Grail quest, Virgil's *Aeneid*, Homer's *Odyssey*, and the *Argonautica*. Cervantes' *Don Quixote* is a parody of chivalric quests. J.R.R. Tolkien's *The Lord of the Rings* trilogy (1954) also features a quest.

One instance of similar themes across cultures is when those cultures are in countries that are geographically close to each other. For example, a folklore story of a rabbit in the moon using a mortar and pestle is shared among China, Japan, Korea, and Thailand—making medicine in China, making rice cakes in Japan and Korea, and hulling rice in Thailand. Another instance is when cultures are more distant geographically, but their languages are related. For example, East Turkestan's Uighurs and people in Turkey share tales of folk hero Effendi Nasreddin Hodja. Another instance, which may either be called cultural diffusion or simply reflect commonalities in the human imagination, involves shared themes among geographically and linguistically different cultures: both Cameroon's and Greece's folklore tell of centaurs; Cameroon, India, Malaysia, Thailand, and Japan, of mermaids; Brazil, Peru, China, Japan, Malaysia, Indonesia, and Cameroon, of underwater civilizations; and China, Japan, Thailand, Vietnam, Malaysia, Brazil, and Peru, of shape-shifters.

Two prevalent literary themes are love and friendship, which can end happily, sadly, or both. William Shakespeare's *Romeo and Juliet,* Emily Brontë's *Wuthering Heights,* Leo Tolstoy's *Anna Karenina,* and both *Pride and Prejudice* and *Sense and Sensibility* by Jane Austen are famous examples. Another theme recurring in popular literature is of revenge, an old theme in dramatic literature, e.g. Elizabethans Thomas Kyd's *The Spanish Tragedy* and Thomas Middleton's *The Revenger's Tragedy.* Some more well-known instances include Shakespeare's tragedies *Hamlet* and *Macbeth,* Alexandre Dumas' *The Count of Monte Cristo,* John Grisham's *A Time to Kill,* and Stieg Larsson's *The Girl Who Kicked the Hornet's Nest.*

Themes are underlying meanings in literature. For example, if a story's main idea is a character succeeding against all odds, the theme is overcoming obstacles. If a story's main idea is one character wanting what another character has, the theme is jealousy. If a story's main idea is a character doing something they were afraid to do, the theme is courage. Themes differ from topics in that a topic is a subject matter; a theme is the author's opinion about it. For example, a work could have a topic of war and a theme that war is a curse. Authors present themes through characters' feelings, thoughts, experiences, dialogue, plot actions, and events. Themes function as "glue" holding other essential story elements together. They offer readers insights into characters' experiences, the author's philosophy, and how the world works.

Using Text Features
Table of Contents and Index
When examining a book, a journal article, a monograph, or other publication, the table of contents is in the front. In books, it is typically found following the title page, publication information (often on the facing side of the title page), and dedication page, when one is included. In shorter publications, the table of contents may follow the title page, or the title on the same page. The table of contents in a book lists the number and title of each chapter and its beginning page number. An index, which is most common in books but may also be included in shorter works, is at the back of the publication. Books, especially academic texts, frequently have two: a subject index and an author index. Readers can look alphabetically for specific subjects in the subject index. Likewise, they can look up specific authors cited, quoted, discussed, or mentioned in the author index.

The index in a book offers particular advantages to students. For example, college course instructors typically assign certain textbooks, but do not expect students to read the entire book from cover to cover immediately. They usually assign specific chapters to read in preparation for specific lectures and/or discussions in certain upcoming classes. Reading portions at a time, some students may find references they either do not fully understand or want to know more about. They can look these topics up in the book's subject index to find them in later chapters. When a text author refers to another author, students can also look up the name in the book's author index to find all page numbers of all other references to that author. College students also typically are assigned research papers to write. A book's subject and author indexes can guide students to pages that may help inform them of other books to use for researching paper topics.

Headings
Headings and subheadings concisely inform readers what each section of a paper contains, as well as showing how its information is organized both visually and verbally. Headings are typically up to about five words long. They are not meant to give in-depth analytical information about the topic of their

section, but rather an idea of its subject matter. Text authors should maintain consistent style across all headings. Readers should not expect headings if there is not material for more than one heading at each level, just as a list is unnecessary for a single item. Subheadings may be a bit longer than headings because they expand upon them. Readers should skim the subheadings in a paper to use them as a map of how content is arranged. Subheadings are in smaller fonts than headings to mirror relative importance. Subheadings are not necessary for every paragraph. They should enhance content, not substitute for topic sentences.

When a heading is brief, simple, and written in the form of a question, it can have the effect of further drawing readers into the text. An effective author will also answer the question in the heading soon in the following text. Question headings and their text answers are particularly helpful for engaging readers with average reading skills. Both headings and subheadings are most effective with more readers when they are obvious, simple, and get to their points immediately. Simple headings attract readers; simple subheadings allow readers a break, during which they also inform reader decisions whether to continue reading or not. Headings stand out from other text through boldface, but also italicizing and underlining them would be excessive. Uppercase-lowercase headings are easier for readers to comprehend than all capitals. More legible fonts are better. Some experts prefer serif fonts in text, but sans-serif fonts in headings. Brief subheadings that preview upcoming chunks of information reach more readers.

Text Features

Textbooks that are designed well employ varied text features for organizing their main ideas, illustrating central concepts, spotlighting significant details, and signaling evidence that supports the ideas and points conveyed. When a textbook uses these features in recurrent patterns that are predictable, it makes it easier for readers to locate information and come up with connections. When readers comprehend how to make use of text features, they will take less time and effort deciphering how the text is organized, leaving them more time and energy for focusing on the actual content in the text. Instructional activities can include not only previewing text through observing main text features, but moreover through examining and deconstructing the text and ascertaining how the text features can aid them in locating and applying text information for learning.

Included among various text features are a table of contents, headings, subheadings, an index, a glossary, a foreword, a preface, paragraphing spaces, bullet lists, footnotes, sidebars, diagrams, graphs, charts, pictures, illustrations, captions, italics, boldface, colors, and symbols. A glossary is a list of key vocabulary words and/or technical terminology and definitions. This helps readers recognize or learn specialized terms used in the text before reading it. A foreword is typically written by someone other than the text author and appears at the beginning to introduce, inform, recommend, and/or praise the work. A preface is often written by the author and also appears at the beginning, to introduce or explain something about the text, like new additions. A sidebar is a box with text and sometimes graphics at the left or right side of a page, typically focusing on a more specific issue, example, or aspect of the subject. Footnotes are additional comments/notes at the bottom of the page, signaled by superscript numbers in the text.

Text Features on Websites

On the Internet or in computer software programs, text features include URLs, home pages, pop-up menus, drop-down menus, bookmarks, buttons, links, navigation bars, text boxes, arrows, symbols, colors, graphics, logos, and abbreviations. URLs (Universal Resource Locators) indicate the internet "address" or location of a website or web page. They often start with www. (world wide web) or http:// (hypertext transfer protocol) or https:// (the "s" indicates a secure site) and appear in the Internet browser's top address bar. Clickable buttons are often links to specific pages on a website or other external sites. Users can click on some buttons to open pop-up or drop-down menus, which offer a list of actions or departments from which to select. Bookmarks are the electronic versions of physical bookmarks. When users bookmark a website/page, a link is established to the site URL and saved, enabling returning to the site in the future without having to remember its name or URL by clicking the bookmark.

Readers can more easily navigate websites and read their information by observing and utilizing their various text features. For example, most fully developed websites include search bars, where users can type in topics, questions, titles, or names to locate specific information within the large amounts stored on many sites. Navigation bars (software developers frequently use the abbreviation term "navbar") are graphical user interfaces (GUIs) that facilitate visiting different sections, departments, or pages within a website, which can be difficult or impossible to find without these. Typically, they appear as a series of links running horizontally across the top of each page. Navigation bars displayed vertically along the left side of the page are also called sidebars. Links, i.e. hyperlinks, enable hyperspeed browsing by allowing readers to jump to new pages/sites. They may be URLs, words, phrases, images, buttons, etc. They are often but not always underlined and/or blue, or other colors.

Analyzing How Authors Construct Arguments

When authors write text for the purpose of persuading others to agree with them, they assume a position with the subject matter about which they are writing. Rather than presenting information objectively, the author treats the subject matter subjectively so that the information presented supports his or her position. In their argumentation, the author presents information that refutes or weakens opposing positions. Another technique authors use in persuasive writing is to anticipate arguments against the position. When students learn to read subjectively, they gain experience with the concept of persuasion in writing, and learn to identify positions taken by authors. This enhances their reading comprehension and develops their skills for identifying pro and con arguments and biases.

There are five main parts of the classical argument that writers employ in a well-designed stance:

- Introduction: In the introduction to a classical argument, the author establishes goodwill and rapport with the reading audience, warms up the readers, and states the thesis or general theme of the argument.

- Narration: In the narration portion, the author gives a summary of pertinent background information, informs the readers of anything they need to know regarding the circumstances and environment surrounding and/or stimulating the argument, and establishes what is at risk or the stakes in the issue or topic. Literature reviews are common examples of narrations in academic writing.

- Confirmation: The confirmation states all claims supporting the thesis and furnishes evidence for each claim, arranging this material in logical order—e.g. from most obvious to most subtle or strongest to weakest.

- Refutation and Concession: The refutation and concession discuss opposing views and anticipate reader objections without weakening the thesis, yet permitting as many oppositions as possible.

- Summation: The summation strengthens the argument while summarizing it, supplying a strong conclusion and showing readers the superiority of the author's solution.

Introduction
A classical argument's introduction must pique reader interest, get readers to perceive the author as a writer, and establish the author's position. Shocking statistics, new ways of restating issues, or quotations

or anecdotes focusing the text can pique reader interest. Personal statements, parallel instances, or analogies can also begin introductions—so can bold thesis statements if the author believes readers will agree. Word choice is also important for establishing author image with readers. The introduction should typically narrow down to a clear, sound thesis statement. If readers cannot locate one sentence in the introduction explicitly stating the writer's position or the point they support, the writer probably has not refined the introduction sufficiently.

Narration and Confirmation

The narration part of a classical argument should create a context for the argument by explaining the issue to which the argument is responding, and by supplying any background information that influences the issue. Readers should understand the issues, alternatives, and stakes in the argument by the end of the narration to enable them to evaluate the author's claims equitably. The confirmation part of the classical argument enables the author to explain why they believe in the argument's thesis. The author builds a chain of reasoning by developing several individual supporting claims and explaining why that evidence supports each claim and also supports the overall thesis of the argument.

Refutation and Concession and Summation

The classical argument is the model for argumentative/persuasive writing, so authors often use it to establish, promote, and defend their positions. In the refutation aspect of the refutation and concession part of the argument, authors disarm reader opposition by anticipating and answering their possible objections, persuading them to accept the author's viewpoint. In the concession aspect, authors can concede those opposing viewpoints with which they agree. This can avoid weakening the author's thesis while establishing reader respect and goodwill for the author: all refutation and no concession can antagonize readers who disagree with the author's position. In the conclusion part of the classical argument, a less skilled writer might simply summarize or restate the thesis and related claims; however, this does not provide the argument with either momentum or closure. More skilled authors revisit the issues and the narration part of the argument, reminding readers of what is at stake.

Evaluating Reasoning and Evidence from Various Sources

Books as Resources

When a student has an assignment to research and write a paper, one of the first steps after determining the topic is to select research sources. The student may begin by conducting an Internet or library search of the topic, may refer to a reading list provided by the instructor, or may use an annotated bibliography of works related to the topic. To evaluate the worth of the book for the research paper, the student first considers the book title to get an idea of its content. Then the student can scan the book's table of contents for chapter titles and topics to get further ideas of their applicability to the topic. The student may also turn to the end of the book to look for an alphabetized index. Most academic textbooks and scholarly works have these; students can look up key topic terms to see how many are included and how many pages are devoted to them.

Journal Articles

Like books, journal articles are primary or secondary sources the student may need to use for researching any topic. To assess whether a journal article will be a useful source for a particular paper topic, a student can first get some idea about the content of the article by reading its title and subtitle, if any exists. Many journal articles, particularly scientific ones, include abstracts. These are brief summaries of the content. The student should read the abstract to get a more specific idea of whether the experiment, literature review, or other work documented is applicable to the paper topic. Students should also check the references at the end of the article, which today often contain links to related works for exploring the topic further.

Encyclopedias and Dictionaries

Dictionaries and encyclopedias are both reference books for looking up information alphabetically. Dictionaries are more exclusively focused on vocabulary words. They include each word's correct spelling, pronunciation, variants, part(s) of speech, definitions of one or more meanings, and examples used in a sentence. Some dictionaries provide illustrations of certain words when these inform the meaning. Some dictionaries also offer synonyms, antonyms, and related words under a word's entry. Encyclopedias, like dictionaries, often provide word pronunciations and definitions. However, they have broader scopes: one can look up entire subjects in encyclopedias, not just words, and find comprehensive, detailed information about historical events, famous people, countries, disciplines of study, and many other things. Dictionaries are for finding word meanings, pronunciations, and spellings; encyclopedias are for finding breadth and depth of information on a variety of topics.

Card Catalogs

A card catalog is a means of organizing, classifying, and locating the large numbers of books found in libraries. Without being able to look up books in library card catalogs, it would be virtually impossible to find them on library shelves. Card catalogs may be on traditional paper cards filed in drawers, or electronic catalogs accessible online; some libraries combine both. Books are shelved by subject area; subjects are coded using formal classification systems—standardized sets of rules for identifying and labeling books by subject and author. These assign each book a call number: a code indicating the classification system, subject, author, and title. Call numbers also function as bookshelf "addresses" where books can be located. Most public libraries use the Dewey Decimal Classification System. Most university, college, and research libraries use the Library of Congress Classification. Nursing students will also encounter the National Institute of Health's National Library of Medicine Classification System, which major collections of health sciences publications utilize.

Databases

A database is a collection of digital information organized for easy access, updating, and management. Users can sort and search databases for information. One way of classifying databases is by content, i.e. full-text, numerical, bibliographical, or images. Another classification method used in computing is by organizational approach. The most common approach is a relational database, which is tabular and defines data so they can be accessed and reorganized in various ways. A distributed database can be reproduced or interspersed among different locations within a network. An object-oriented database is organized to be aligned with object classes and subclasses defining the data. Databases usually collect files like product inventories, catalogs, customer profiles, sales transactions, student bodies, and resources. An associated set of application programs is a database management system or database manager. It enables users to specify which reports to generate, control access to reading and writing data, and analyze database usage. Structured Query Language (SQL) is a standard computer language for updating, querying, and otherwise interfacing with databases.

Identifying Primary Sources in Various Media

A primary source is a piece of original work. This can include books, musical compositions, recordings, movies, works of visual art (paintings, drawings, photographs), jewelry, pottery, clothing, furniture, and other artifacts. Within books, primary sources may be of any genre. Whether nonfiction based on actual events or a fictional creation, the primary source relates the author's firsthand view of some specific event, phenomenon, character, place, process, ideas, field of study or discipline, or other subject matter. Whereas primary sources are original treatments of their subjects, secondary sources are a step removed from the original subjects; they analyze and interpret primary sources. These include journal articles, newspaper or magazine articles, works of literary criticism, political commentaries, and academic textbooks.

In the field of history, primary sources frequently include documents that were created around the same time period that they were describing, and most often produced by someone who had direct experience or knowledge of the subject matter. In contrast, secondary sources present the ideas and viewpoints of other authors about the primary sources; in history, for example, these can include books and other written works about the particular historical periods or eras in which the primary sources were produced. Primary sources pertinent in history include diaries, letters, statistics, government information, and original journal articles and books. In literature, a primary source might be a literary novel, a poem or book of poems, or a play. Secondary sources addressing primary sources may be criticism, dissertations, theses, and journal articles. Tertiary sources, typically reference works referring to primary and secondary sources, include encyclopedias, bibliographies, handbooks, abstracts, and periodical indexes.

In scientific fields, when scientists conduct laboratory experiments to answer specific research questions and test hypotheses, lab reports and reports of research results constitute examples of primary sources. When researchers produce statistics to support or refute hypotheses, those statistics are primary sources. When a scientist is studying some subject longitudinally or conducting a case study, they may keep a journal or diary. For example, Charles Darwin kept diaries of extensive notes on his studies during sea voyages on the *Beagle*, visits to the Galápagos Islands, etc.; Jean Piaget kept journals of observational notes for case studies of children's learning behaviors. Many scientists, particularly in past centuries, shared and discussed discoveries, questions, and ideas with colleagues through letters, which also constitute primary sources. When a scientist seeks to replicate another's experiment, the reported results, analysis, and commentary on the original work is a secondary source, as is a student's dissertation if it analyzes or discusses others' work rather than reporting original research or ideas.

Practice Questions

Passage I: Literature

The following passage is taken from Chapter 6 of Sense and Sensibility by Jane Austen:

The first part of their journey was performed in too melancholy a disposition to be otherwise than tedious and unpleasant. But as they drew toward the end of it, their interest in the appearance of a country which they were to inhabit overcame their dejection, and a view of Barton Valley as they entered it gave them cheerfulness. It was a pleasant fertile spot, well wooded, and rich in pasture. After winding along it for more than a mile, they reached their own house. A small green court was the whole of its demesne in front; and a neat wicket gate admitted them into it.

As a house, Barton Cottage, though small, was comfortable and compact; but as a cottage it was defective, for the building was regular, the roof was tiled, the window shutters were not painted green, nor were the walls covered with honeysuckles. A narrow passage led directly through the house into the garden behind. On each side of the entrance was a sitting room, about sixteen feet square; and beyond them were the offices and the stairs. Four bed-rooms and two garrets formed the rest of the house. It had not been built many years and was in good repair. In comparison of Norland, it was poor and small indeed!—but the tears which recollection called forth as they entered the house were soon dried away. They were cheered by the joy of the servants on their arrival, and each for the sake of the others resolved to appear happy. It was very early in September; the season was fine, and from first seeing the place under the advantage of good weather, they received an impression in its favour which was of material service in recommending it to their lasting approbation.

The situation of the house was good. High hills rose immediately behind, and at no great distance on each side; some of which were open downs, the others cultivated and woody. The village of Barton was chiefly on one of these hills, and formed a pleasant view from the cottage windows. The prospect in front was more extensive; it commanded the whole of the valley, and reached into the country beyond. The hills which surrounded the cottage terminated the valley in that direction; under another name, and in another course, it branched out again between two of the steepest of them.

With the size and furniture of the house Mrs. Dashwood was upon the whole well satisfied; for though her former style of life rendered many additions to the latter indispensable, yet to add and improve was a delight to her; and she had at this time ready money enough to supply all that was wanted of greater elegance to the apartments. "As for the house itself, to be sure," said she, "it is too small for our family, but we will make ourselves tolerably comfortable for the present, as it is too late in the year for improvements. Perhaps in the spring, if I have plenty of money, as I dare say I shall, we may think about building. These parlors are both too small for such parties of our friends as I hope to see often collected here; and I have some thoughts of throwing the passage into one of them with perhaps a part of the other, and so leave the remainder of that other for an entrance; this, with a new drawing room which may be easily added, and a bed-chamber and garret above, will make it a very snug little cottage. I could wish the stairs were handsome. But one must not expect every thing; though I suppose it would be no difficult matter to widen them. I shall see how much I am before-hand with the world in the spring, and we will plan our improvements accordingly."

In the mean time, till all these alterations could be made from the savings of an income of five hundred a-year by a woman who never saved in her life, they were wise enough to be contented with the house as it was; and each of them was busy in arranging their particular concerns, and endeavoring, by placing around them books and other possessions, to form themselves a home. Marianne's pianoforte was unpacked and properly disposed of; and Elinor's drawings were affixed to the walls of their sitting room.

In such employments as these they were interrupted soon after breakfast the next day by the entrance of their landlord, who called to welcome them to Barton, and to offer them every accommodation from his own house and garden in which theirs might at present be deficient. Sir John Middleton was a good looking man about forty. He had formerly visited at Stanhill, but it was too long for his young cousins to remember him. His countenance was thoroughly good-humoured; and his manners were as friendly as the style of his letter. Their arrival seemed to afford him real satisfaction, and their comfort to be an object of real solicitude to him. He said much of his earnest desire of their living in the most sociable terms with his family, and pressed them so cordially to dine at Barton Park every day till they were better settled at home, that, though his entreaties were carried to a point of perseverance beyond civility, they could not give offence. His kindness was not confined to words; for within an hour after he left them, a large basket full of garden stuff and fruit arrived from the park, which was followed before the end of the day by a present of game.

1. What is the point of view in this passage?
   a. Third-person omniscient
   b. Second-person
   c. First-person
   d. Third-person objective

2. Which of the following events occurred first?
   a. Sir John Middleton stopped by for a visit.
   b. The servants joyfully cheered for the family.
   c. Mrs. Dashwood discussed improvements to the cottage.
   d. Elinor hung her drawings up in the sitting room.

3. Over the course of the passage, the Dashwoods' attitude shifts. Which statement best describes that shift?
   a. From appreciation of the family's former life of privilege to disdain for the family's new landlord
   b. From confidence in the power of the family's wealth to doubt in the family's ability to survive
   c. From melancholy about leaving Norland to excitement about reaching Barton Cottage in the English countryside
   d. From cheerfulness about the family's expedition to anxiety about the upkeep of such a big home

4. Which of the following is a theme of this passage?
   a. All-conquering love
   b. Power of wealth
   c. Wisdom of experience
   d. Reality vs. expectations

5. At the start of paragraph five, the narrator says, "till all these alterations could be made from the savings of an income of five hundred a-year by a woman who never saved in her life, they were wise enough to be contented with the house as it was." What does the narrator mean?

    a. The family is going through a transition phase.

    b. Mrs. Dashwood needs to obtain meaningful employment.

    c. The family is going through a growth phase.

    d. The Dashwood children need to be concerned about the future.

6. What is the relationship between the new landlord and the Dashwoods?

    a. He is a former social acquaintance.

    b. He is one of their cousins.

    c. He is Mrs. Dashwood's father.

    d. He is a long-time friend of the family.

7. Why does the narrator describe the generosity of Sir John Middleton?

    a. To identify one of many positive traits that a landlord should possess.

    b. To explain how a landlord should conduct himself in order to be successful.

    c. To illustrate how his kindness eased the family's adaptation to their new home and circumstances.

    d. To demonstrate that he did not need to be cold and businesslike all of the time.

8. At the start of paragraph two, the narrator refers to the Dashwoods' new home as being "defective." What does the narrator mean?

    a. The tall hills surround and hide the home from neighboring structures.

    b. The building is much too poor and crowded for the family.

    c. The home's insufficiently sized parlors are too small for entertaining.

    d. The building's look and feel do not resemble that of a typical cottage.

9. Which of the following best describes the tone of the passage?

    a. Melancholy

    b. Inventive

    c. Upbeat

    d. Apprehensive

10. Toward the end of paragraph six, the narrator says, "his entreaties were carried to a point of perseverance beyond civility." What does the narrator accomplish by saying this?

    a. Signifies the Dashwoods' annoyance with the fake friendliness of Sir John Middleton

    b. Describes Sir John Middleton's slightly overbearing, overly effusive invitations

    c. Demonstrates the language from the time period in which the piece was penned

    d. Questions the genuineness of the offer for the family to stay at Barton Cottage

Passage II: Social Science

The following passage is taken from the Advantages of Division of Labor section in Chapter VI of <u>Principles of Political Economy</u> *by John Stuart Mill:*

The causes of the increased efficiency given to labor by the division of employments are some of them too familiar to require specification; but it is worthwhile to attempt a complete enumeration of them. By Adam Smith they are reduced to three: "First, the increase of dexterity in every particular workman; secondly, the saving of the time which is commonly lost in passing from one species of work to another; and, lastly, the invention of a great number of machines which facilitate and abridge labor, and enable one man to do the work of many."

Of these, the increase of dexterity of the individual workman is the most obvious and universal. It does not follow that because a thing has been done oftener it will be done better. That depends on the intelligence of the workman, and on the degree in which his mind works along with his hands. But it will be done more easily. This is as true of mental operations as of bodily. Even a child, after much practice, sums up a column of figures with a rapidity which resembles intuition. The act of speaking any language, of reading fluently, of playing music at sight, are cases as remarkable as they are familiar. Among bodily acts, dancing, gymnastic exercises, ease and brilliancy of execution on a musical instrument, are examples of the rapidity and facility acquired by repetition. In simpler manual operations the effect is, of course, still sooner produced.

The second advantage enumerated by Adam Smith as arising from the division of labor is one on which I can not help thinking that more stress is laid by him and others than it deserves. To do full justice to his opinion, I will quote his own exposition of it: "It is impossible to pass very quickly from one kind of work to another, that is carried on in a different place, and with quite different tools. A country weaver, who cultivates a small farm, must lose a good deal of time in passing from his loom to the field, and from the field to his loom. When the two trades can be carried on in the same workhouse, the loss of time is no doubt much less. It is even in this case, however, very considerable. A man commonly saunters a little in turning his hand from one sort of employment to another." I am very far from implying that these considerations are of no weight; but I think there are counter-considerations which are overlooked. If one kind of muscular or mental labor is different from another, for that very reason it is to some extent a rest from that other; and if the greatest vigor is not at once obtained in the second occupation, neither could the first have been indefinitely prolonged without some relaxation of energy. It is a matter of common experience that a change of occupation will often afford relief where complete repose would otherwise be necessary, and that a person can work many more hours without fatigue at a succession of occupations, than if confined during the whole time to one. Different occupations employ different muscles, or different energies of the mind, some of which rest and are refreshed while others work. Bodily labor itself rests from mental, and conversely. The variety itself has an invigorating effect on what, for want of a more philosophical appellation, we must term the animal spirits—so important to the efficiency of all work not mechanical, and not unimportant even to that.

The third advantage attributed by Adam Smith to the division of labor is, to a certain extent, real. Inventions tending to save labor in a particular operation are more likely to occur to any one in proportion as his thoughts are intensely directed to that occupation, and continually employed upon it.

This also can not be wholly true. "The founder of the cotton manufacture was a barber. The inventor of the power-loom was a clergyman. A farmer devised the application of the screw-propeller. A fancy-goods shopkeeper is one of the most enterprising experimentalists in agriculture. The most remarkable architectural design of our day has been furnished by a gardener. The first person who supplied London with water was a goldsmith. The first extensive maker of English roads was a blind man, bred to no trade. The father of English inland navigation was a duke, and his engineer was a millwright. The first great builder of iron bridges was a stone-mason, and the greatest railway engineer commenced his life as a colliery engineer."

(4.) The greatest advantage (next to the dexterity of the workmen) derived from the minute division of labor which takes place in modern manufacturing industry, is one not mentioned by Adam Smith, but to which attention has been drawn by Mr. Babbage: the more economical distribution of labor by classing the work-people according to their capacity.

11. Which of the following statements would the author agree is an advantage associated with the division of labor proposed by Adam Smith?
    a. Dexterity increases as employees complete repeated tasks
    b. Repetition results in increased monotony for employees
    c. Greater interdependence forms in the production process
    d. Unemployment increases as workers are replaced by machines

12. In paragraph five, why does the author mention that the "first great builder of iron bridges was a stone-mason"?
    a. To provide an example of how a person can do anything that he or she sets his or her mind to
    b. To demonstrate that Adam Smith's third advantage associated with the division of labor is not entirely true
    c. To explain that individuals are not always employed in professions for which they have received schooling
    d. To state that it is possible for individuals to have more than one career during their working life

13. Which statement best expresses the passage's main idea?
    a. The effect of specialization of division of labor
    b. Disproving the economic principles of Adam Smith
    c. Advantages associated with the division of labor
    d. Basic principles of macroeconomics

14. Which advantage associated with the division of labor does the author say is the second most important one?
    a. Increase of dexterity in employees
    b. Savings of time by staying on one type of work
    c. Invention of machines to assist with manual labor
    d. Classification of employees by their abilities

15. In paragraph two of this passage, what does the word "bodily" mean?
    a. Tangible
    b. Animal
    c. Organic
    d. Spiritual

16. Which word best describes the author's attitude toward Adam Smith?
    a. Dismissive
    b. Respectful
    c. Adoring
    d. Questioning

17. Based on the second advantage mentioned in the passage, what would be the most efficient course of action for an employee working in a copy center?
    a. The employee should focus on only a single task, such as producing black and white copies for large company orders.
    b. The employee should produce copies in one room and then go to a different building hall to collate and bind documents.
    c. The employee should produce black and white copies for a few hours and then change their focus to shipping out copy orders to customers.
    d. The employee should be allowed unlimited flexibility to schedule their daily work tasks however they see fit.

18. Which idea is supported by this passage?
    a. Repetitive tasks lead to stability in employees' work.
    b. One employee should complete all of the tasks in a workflow.
    c. Electronic toll collection is taking jobs from workers.
    d. Milking machines help to speed up the retrieval of milk from dairy farms.

19. A doll manufacturer opens a new facility and needs employees who can cut fabric, stuff bodies, and make delicate hair and facial features. Instead of hiring employees and cross-training them on all the tasks, the manufacturer hires staff with varying degrees of skill for the various tasks. What is this is an example of?
    a. Classing employees by their capacity
    b. Increasing dexterity by training on repetitive tasks
    c. Exclusively hiring experienced employees
    d. Introducing machines to assist with unfamiliar tasks

20. In paragraph three of this passage, why does the author use the phrase "animal spirits"?
    a. To provide an example of how employees feel when they work beyond fatigue
    b. To explain the positive effect that variety in work has on employees
    c. To demonstrate how employees feel when they do not receive proper training
    d. To state how employees are aversely affected when they change tasks

Passage III (A and B): Humanities

Passage A
The following passage is taken from Compassion and Benevolence under The Affections in Section II Of The Philosophy of the Moral Feelings by John Abercrombie, M.D. OXON. & EDIN:

> The exercise of the benevolent affections may be briefly treated of, under nearly the same heads as those referred to when considering the principle of Justice;—keeping in mind that they lead to greater exertion for the benefit of others, and thus often demand a greater sacrifice of self-love, than is included under the mere requirements of justice. On the other hand, benevolence is not to be exercised at the expense of Justice; as would be the case, if a man were found relieving

distress by such expedients as involve the necessity of withholding the payment of just debts, or imply the neglect or infringement of some duty which he owes to another.

Compassion and benevolent exertion are due toward alleviating the distresses of others. This exercise of them, in many instances, calls for a decided sacrifice of personal interest, and, in others, for considerable personal exertion. We feel our way to the proper measure of these sacrifices, by the high principle of moral duty, along with that mental exercise which places us in the situation of others, and, by a kind of reflected self-love, judges of the conduct due by us to them in our respective circumstances.—The details of this subject would lead us into a field too extensive for our present purpose. Pecuniary aid, by those who have the means, is the most easy form in which benevolence can be gratified, and that which often requires the least, if any, sacrifice of personal comfort or self-love. The same affection maybe exercised in a degree much higher in itself, and often much more useful to others, by personal exertion and personal kindness. The former, compared with the means of the individual, may present a mere mockery of mercy; while the latter, even in the lowest walks of life, often exhibit the brightest displays of active usefulness that can adorn the human character. This high and pure benevolence not only is dispensed with willingness, when occasions present themselves; but seeks out opportunities for itself, and feels in want of its natural and healthy exercise when deprived of an object on which it may be bestowed.

Benevolence is to be exercised toward the reputation of others. This consists not only in avoiding any injury to their characters, but in exertions to protect them against the injustice of others,—to correct misrepresentations,—to check the course of slander, and to obviate the efforts of those who would poison the confidence of friends, or disturb the harmony of society.

Passage B
The following passage is taken from Part II in Section II Of Benevolence in An Enquiry Concerning the Principles of Morals, by David Hume:

Giving alms to common beggars is naturally praised; because it seems to carry relief to the distressed and indigent: but when we observe the encouragement thence arising to idleness and debauchery, we regard that species of charity rather as a weakness than a virtue.

Tyrannicide, or the assassination of usurpers and oppressive princes, was highly extolled in ancient times; because it both freed mankind from many of these monsters, and seemed to keep the others in awe, whom the sword or poignard could not reach. But history and experience having since convinced us, that this practice increases the jealousy and cruelty of princes, a Timoleon and a Brutus, though treated with indulgence on account of the prejudices of their times, are now considered as very improper models for imitation.

Liberality in princes is regarded as a mark of beneficence, but when it occurs, that the homely bread of the honest and industrious is often thereby converted into delicious cates for the idle and the prodigal, we soon retract our heedless praises. The regrets of a prince, for having lost a day, were noble and generous: but had he intended to have spent it in acts of generosity to his greedy courtiers, it was better lost than misemployed after that manner.

Luxury, or a refinement on the pleasures and conveniences of life, had not long been supposed the source of every corruption in government, and the immediate cause of faction, sedition, civil wars, and the total loss of liberty. It was, therefore, universally regarded as a vice, and was an object of declamation to all satirists, and severe moralists. Those, who prove, or attempt to prove, that such refinements rather tend to the increase of industry, civility, and arts regulate anew our moral as well as political sentiments, and represent, as laudable or innocent, what had formerly been regarded as pernicious and blamable.

Upon the whole, then, it seems undeniable, that nothing can bestow more merit on any human creature than the sentiment of benevolence in an eminent degree; and that a part, at least, of its merit arises from its tendency to promote the interests of our species, and bestow happiness on human society. We carry our view into the salutary consequences of such a character and disposition; and whatever has so benign an influence, and forwards so desirable an end, is beheld with complacency and pleasure. The social virtues are never regarded without their beneficial tendencies, nor viewed as barren and unfruitful. The happiness of mankind, the order of society, the harmony of families, the mutual support of friends, are always considered as the result of their gentle dominion over the breasts of men.

21. According to the author of passage A, which example of benevolence is the simplest to execute?
   a. Providing money to enable a student the opportunity to attend an educational workshop
   b. Holding a lemonade stand to raise funds and awareness for pediatric cancer
   c. Shopping for new toys to donate to a fundraiser that collects gifts for needy kids
   d. Volunteering to cook and serve Thanksgiving dinner for homeless people

22. In paragraph two of passage A, what kind of "mental exercise" does the author discuss?
   a. Practicing to improve an individual's recall from memory
   b. Trying to see things from another person's point of view
   c. Strengthening a person's ability to concentrate and focus
   d. Taking only a singular perspective into consideration

23. In paragraph one of passage A, what does the term "heads" mean?
   a. Discord
   b. Unlikeness
   c. Manner
   d. Opposition

24. According to passage A, which statement accurately reflects the relationship between benevolence and justice?
   a. Benevolence can be exercised at the expense of justice
   b. Acts of justice require selflessness
   c. Benevolence can be offered in lieu of payment of debts
   d. Justice can be exercised at the expense of benevolence

25. Why does the author of passage B say that giving food and money to beggars is seen as a weakness?
   a. That type of charity encourages laziness and corruption
   b. Some individuals are not truly deserving of the charity
   c. The grants cannot reach all who are affected by poverty
   d. Individuals refuse to accept the handouts due to their pride

26. In paragraph five of passage B, the author uses the phrase "their gentle dominion over the breasts of men." What has dominion over the breasts of men, according to the author?
    a. Merit
    b. The sentiment of benevolence
    c. The social virtues
    d. Complacency and pleasure

27. Which word best describes the author's attitude toward luxury in passage B?
    a. Adoring
    b. Nostalgic
    c. Objective
    d. Dismissive

28. Which statement best describes the way the two passages use point of view?
    a. Passage A is written in second-person point of view, and passage B is written in third-person objective point of view.
    b. Passage A is written in first-person point of view, and passage B is written in third-person objective point of view.
    c. Passage A is written in third-person objective point of view, and passage B is written in third-person limited omniscient point of view.
    d. Passage A is written in first-person point of view, and passage B is written in first-person point of view.

29. Which of the following statements best explains the difference in the tones of the passages?
    a. The tone of passage A is objective, and the tone of passage B is earnest.
    b. The tone of passage A is cynical, and the tone of passage B is sarcastic.
    c. The tone of passage A is excited, and the tone of passage B is ambivalent.
    d. The tone of passage A is sarcastic, and the tone of passage B is cynical.

30. The author of passage B stresses the importance of exercising benevolence for which purpose?
    a. Reducing the suffering of another person
    b. Contributing to the overall happiness of society
    c. Preventing damage to someone else's good name
    d. Attracting good karma to oneself and others

Passage IV: Natural Science

The following passage is taken from the Muscles, Tendons, and Tendon Sheaths section in Chapter XVIII of Manual of Surgery by Alexis Thomson, F.R.C.S. Ed. and Alexander Miles, F.R.C.S. Ed:

Tendon sheaths have the same structure and function as the synovial membranes of joints and are liable to the same diseases. Apart from the tendon sheaths displayed in anatomical dissections, there is a loose peritendinous and perimuscular cellular tissue that is subject to the same pathological conditions as the tendon sheaths proper.

Tenosynovitis. The toxic or infective agent is conveyed to the tendon sheaths through the blood-stream, as in the gouty, gonorrheal, and tuberculous varieties, or is introduced directly through a wound, as in the common pyogenic form of tenosynovitis.

Tenosynovitis crepitans: In the simple or traumatic form of tenosynovitis, although the most prominent etiological factor is a strain or overuse of the tendon, there would appear to be some other, probably a toxic, factor in its production; otherwise the affliction would be much more common than it is: only a small proportion of those who strain or overuse their tendons become the subjects of tenosynovitis. The opposed surfaces of the tendon and its sheath are covered with fibrinous lymph, so that there is friction when they move on one another.

The *clinical features* are pain on movement, tenderness on pressure over the affected tendon, and a sensation of crepitation or friction when the tendon is moved in its sheath. The crepitation may be soft like the friction of snow, or may resemble the creaking of new leather—"saddle-back creaking." There may be swelling in the long axis of the tendon, and redness and edema of the skin. If there is an effusion of fluid into the sheath, the swelling is more marked and crepitation is absent. There is little tendency to the formation of adhesions.

In the upper extremity, the sheath of the long tendon of the biceps may be affected, but the condition is most common in the tendons about the wrist, particularly in the extensors of the thumb, and it is most frequently met with in those who follow occupations which involve prolonged use or excessive straining of these tendons—for example, washerwomen or riveters. It also occurs as a result of excessive piano-playing, fencing, or rowing.

At the ankle it affects the peronei, the extensor digitorum longus, or the tibialis anterior. It is most often met with in relation to the tendo-calcaneus—*Achillo-dynia*—and results from the pressure of ill-fitting boots or from the excessive use and strain of the tendon in cycling, walking, or dancing. There is pain in raising the heel from the ground, and creaking can be felt on palpation.

The *treatment* consists in putting the affected tendon at rest, and with this object a splint may be helpful; the usual remedies for inflammation are indicated: Bier's hyperemia, lead and opium fomentations, and ichthyol and glycerin. The affliction readily subsides under treatment, but is liable to relapse on a repetition of the exciting cause.

Gouty tenosynovitis: A deposit of urate of soda beneath the endothelial covering of tendons or of that lining their sheaths is commonly met with in gouty subjects. The accumulation of urates may result in the formation of visible nodular swellings, varying in size from a pea to a cherry, attached to the tendon and moving with it. They may be merely unsightly, or they may interfere with the use of the tendon. Recurrent attacks of inflammation are prone to occur. We have removed such gouty masses with satisfactory results.

Suppurative tenosynovitis: This form usually follows upon infected wounds of the fingers—especially of the thumb or little finger—and is a frequent sequel to whitlow; it may also follow amputation of a finger. Once the infection has gained access to the sheath, it tends to spread, and may reach the palm or even the forearm, being then associated with cellulitis. In moderately acute cases the tendon and its sheath become covered with granulations, which subsequently lead to the formation of adhesions; while in more acute cases the tendon sloughs. The pus may burst into the cellular tissue outside the sheath, and the suppuration is liable to spread to neighbouring sheaths or to adjacent bones or joints—for example, those of the wrist.

The *treatment* consists in inducing hyperemia and making small incisions for the escape of pus. The site of incision is determined by the point of greatest tenderness on pressure. After the inflammation has subsided, active and passive movements are employed to prevent the formation of adhesions between the tendon and its sheath. If the tendon sloughs, the dead portion should be cut away, as its separation is extremely slow and is attended with prolonged suppuration.

31. Which statement best expresses the main idea of the passage?
    a. A discussion of the causes, symptoms, and treatments associated with various types of tenosynovitis
    b. The similarities that exist between tendon sheaths and the synovial membranes of joints
    c. An exploration of sports and professions that may be responsible for injuries to tendons
    d. Differences in how tenosynovitis displays in injuries of the wrist versus injuries of the ankle

32. According to the passage, which of the following is a treatment for tenosynovitis crepitans of the ankle?
    a. Depositing urate of soda under the endothelial tendon covering
    b. Placing the tendon at rest, possibly utilizing a splinting device
    c. Incorporating active and passive movements to prevent tendon adhesions
    d. Inducing hyperemia and making small incisions to release pus

33. Which statement is supported by the information in the third paragraph?
    a. Tenosynovitis crepitans directly results from the overuse of a tendon.
    b. Edema and redness of the skin will return a diagnosis of tenosynovitis crepitans.
    c. Since few overuse injuries result in tenosynovitis crepitans, a toxic agent may be at play.
    d. A dancer who strains a tendon will undoubtedly suffer from tenosynovitis crepitans.

34. According to the author, a patient having which of the following clinical features would be presenting with a case of Tenosynovitis crepitans?
    a. Visible nodular swellings
    b. Cellulitis infections
    c. Finger wound infections
    d. Sensation of friction

35. The authors' point of view in this passage can be described as which of the following?
    a. First-person
    b. Second-person
    c. Third-person limited omniscient
    d. Third-person objective

36. According to the passage, which of the following may be the end result of an infected wound of the finger?
    a. The formation of cellulitis
    b. Pus spreading to adjacent bones and joints of the wrist
    c. Amputation of the finger
    d. Formation of granules on tendons and sheaths

37. In paragraph four, the authors use the phrases "soft like the friction of snow" and "resemble the creaking of new leather" to describe which of the following?
    a. Pain and pressure due to an increase in movement
    b. Swelling in the long axis of the tendon
    c. Friction associated with tendon movement in the sheath
    d. Effusion of fluid in the sheath of the tendon

38. A construction worker is wearing work boots that are too tight, and he experiences pain when he lifts his heel. Based on the passage, he will most likely be diagnosed with which of the following?
  a. Tenosynovitis
  b. Suppurative tenosynovitis
  c. Gouty tenosynovitis
  d. Tenosynovitis crepitans

39. Why do the authors of this passage incorporate the words "washerwomen" and "riveters" in paragraph five?
  a. To give examples of repetitive motion jobs that may result in tenosynovitis crepitans
  b. To emphasize the importance of healthcare for blue-collar workers
  c. To provide illustrations of recurrent inflammation attacks common to gouty tenosynovitis
  d. To illustrate why ergonomics in the workplace is essential to the health of employees

40. A woman presents to her physician's office with a nodular swelling on her thumb the size of a blueberry. Based on the passage, which condition will she most likely be diagnosed with?
  a. Tendo-calcaneus
  b. Gouty tenosynovitis
  c. Suppurative tenosynovitis
  d. Tenosynovitis

Answer Explanations

1. A: The point of view of the narrator of the passage can best be described as third-person omniscient. The narrator refers to the characters in the story by third-person pronouns: *he, it,* and *they.* The narrator also comes across as "all-knowing" (omniscient) by relating the information and feelings about all the characters (instead of just those of a single character). Second-person point of view would incorporate the second-person pronoun, *you.* First-person point of view would incorporate first-person pronouns: *I* and *we.* Finally, the third-person objective point of view would also refer to the characters in the story by third-person pronouns *he, it,* and *they.* However, the narrator would stay detached, only telling the story and not expressing what the characters feel and think.

2. B: The chronological order of these four events in the passage is as follows:

1. The servants joyfully cheered for the family.
2. Mrs. Dashwood discussed improvements to the cottage.
3. Elinor hung her drawings up in the sitting room.
4. Sir John Middleton stopped by for a visit.

3. C: In the course of this passage, the Dashwoods' attitude shifts. The narrator initially describes the family's melancholy disposition at the start of their journey from Norland. However, the narrator then describes their spirits beginning to lift and becoming more upbeat as they make their way through the scenic English countryside to their new home at Barton Cottage. Although the remaining answer choices may contain some partial truths (appreciation for the family's former privileged life and fear about the family's expedition), they do not accurately depict the narrator's theme throughout the entire passage. For example, the family did not express disdain for their new landlord or doubt their future ability to survive.

4. D: Reality versus expectations is a theme of this passage. Although Mrs. Dashwood expects her new financial circumstances to be trying, and the family expects to miss their former place of residence for quite some time, the Dashwoods seem to begin to adapt well with the help of Sir John Middleton, their generous new landlord. A theme of love conquers all is typically used in literature when a character overcomes an obstacle due to his or her love for someone. A theme of power of wealth is used in literature to show that money either accomplishes things or is the root of evil. Finally, a theme of wisdom of experience is typically used in literature to show that improved judgment comes with age.

5. A: The statement at the start of paragraph five in this passage signifies the family is going through a transition. The Dashwoods find themselves in a time of great transition as they learn to accept their family's demotion in social standing and their reduced income. The statement was not meant to signify that Mrs. Dashwood needs to obtain meaningful employment, that the family is going through a growth phase, or that the Dashwood children need to be concerned about the future.

6. B: The new landlord is a cousin to the Dashwoods. This accurately depicts their relationship. The new landlord was not a former social acquaintance, Mrs. Dashwood's father, or a long-time friend of the family.

7. C: The narrator describes the generosity of Sir John Middleton to explain how his kindness eased the family's adaptation to their new home and circumstances. He played a very important role in making the Dashwoods feel comfortable moving to Barton Cottage. The narrator did not describe Sir John Middleton's generosity to identify one of the many positive traits that a landlord should possess, to explain how a landlord should conduct himself in order to be successful, or to demonstrate that he did not need to be cold and businesslike all the time.

8. D: The narrator's reference to the Dashwood's new home as being "defective" means that the building's look and feel do not resemble that of a typical cottage. The narrator describes the building as being regular, with a tiled roof (rather than a thatched roof), with shutters that are not green, and with walls not covered with honeysuckles—all characteristics that a traditional cottage would possess. Although the other three answer choices are true statements about their new home and property, they do not relate to the narrator's statement that the new home is "defective."

9. C: The overall tone of the passage is upbeat. Although the passage begins with the family in a melancholy disposition at the start of their journey (and perhaps a bit apprehensive), their spirits begin to lift as they make their way through the scenic English countryside to Barton Cottage.

10. B: The narrator is describing how exuberantly Sir John Middleton expresses his pleasure at gaining the Dashwoods as tenants. Sir John Middleton is overjoyed to welcome his cousins to the cottage on his property, which he expresses through his words, gifts, and invitations to dine with him. He just goes a bit too far in demonstrating his delight, for which the Dashwoods forgive him. Sir John Middleton does not express any fake friendliness. The phrase was not used to demonstrate the language from the time period in which the piece was penned. Finally, Sir John Middleton's offer for the Dashwoods to stay at Barton Cottage is indeed genuine.

11. A: The author would agree that an advantage associated with the division of labor proposed by Adam Smith is that dexterity increases as employees complete repeated tasks. This answer can be inferred from the second paragraph. The remaining three answer choices are known disadvantages related to the division of labor concept.

12. B: The author mentions that the "first great builder of iron bridges was a stone-mason" in paragraph five in order to demonstrate how the third advantage associated with the division of labor proposed by Adam Smith is not entirely true. Even though the author agrees with Adam Smith that the invention of machines helps employees to save time, he does not agree that employees who are intimately involved in a daily routine of work come up with the ideas for those inventions (e.g., "the inventor of the power-loom was a clergyman," not a weaver). The remaining answer choices are not addressed by the author in the passage.

13. C: The passage's main idea is best expressed by the answer choice "advantages associated with the division of labor." The passage does not focus solely on the effect of specialization of labor, which consists of increasing productivity by dividing up larger tasks into smaller tasks to be completed by workers with specialized skills. The passage is not concerned with disproving Adam Smith's economic principles. Finally, macroeconomics deals with the larger economy as a whole and topics such as interest rates and gross domestic product. Thus, macroeconomics is not discussed in this passage.

14. D: In paragraph six, the author lists classification of employees by their abilities as the second most important advantage associated with the division of labor. The author states that this is "the greatest advantage (next to the dexterity of the workmen)," which is not mentioned as an advantage by Adam Smith. The remaining three answer choices are listed in the passage as advantages associated with the division of labor. However, they are not listed by the author as being the second most important advantage.

15. A: The author uses the word "bodily" in the second paragraph to mean tangible. The author is referring to the hands-on acts of dancing, gymnastics, and playing musical instruments. "Animal" as an adjective relates to the natural state or primal nature of something, while "organic" similarly refers to something natural. These words are similar to "bodily" but do not fit the author's meaning in this passage, so Choices *B* and *C* are incorrect. "Spiritual" is an antonym to "bodily," so Choice *D* is incorrect.

16. B: The author's attitude toward Adam Smith in this passage can be best described as respectful. The author does not adore Adam Smith, nor does he question or dismiss him and his theory. However, the author expands on the advantages associated with the division of labor to add in his own points where appropriate, and he describes one additional advantage not mentioned by Adam Smith.

17. C: Based on the second advantage, the author would suggest the most efficient course of action for an employee working in a copy center would be to produce black and white copies for a few hours and then change their focus to shipping out copy orders to customers. This is where the author says that he differs from Adam Smith. He believes that there is a benefit to having an employee gain relief from experiencing a change of occupation (or tasks) throughout their work day. Therefore, the author would not suggest that the most efficient course of action is having an employee focus on only a single task (as Adam Smith believed). Completing tasks in different locations is an idea that both the author and Adam Smith would not agree with. They were both proponents of performing tasks in one same physical location to cut down on transfer time. Finally, the author does not mention granting an employee unlimited flexibility in scheduling their work tasks.

18. D: This passage supports the idea of milking machines being used to help speed up the retrieval of milk from dairy farms. This is related to the third advantage: inventions tend to save labor. The passage does not mention that repetitive tasks may lead to stability in employees' work. Although the author mentions there is a benefit to having an employee gain relief from experiencing a change of occupation (or tasks) throughout their work day, he does not state that one employee should complete all the tasks in a workflow. Finally, the passage does not mention the downside of inventions, which is that sometimes they result in a loss of employee jobs.

19. A: Staffing the doll factory is an example of classing employees by their capacity; staff will be hired based on their skills for the tasks that are involved with manufacturing dolls. There is no mention of training employees on repetitive tasks to increase their dexterity or exclusively hiring only experienced employees. Finally, there is also no mention of introducing machines to assist employees with unfamiliar tasks.

20. B: The author uses the phrase "animal spirits" in paragraph three to explain the positive effect that variety in work has on employees. This is one of the areas in which the author expands on one of Adam Smith's advantages of the division of labor. Adam Smith says that time can be saved if employees focus on one type of work. The author is not dismissing his claim. However, he mentions that some employees may instead benefit from changing the type of work that they do during the day, and such variety invigorates some employees' animal spirits. The remaining answer choices (working beyond fatigue, failing to receive proper training, and employees being adversely affected when they change tasks) are not mentioned in the passage.

21. A: Providing money to allow a student the opportunity to attend an educational workshop is the simplest way to do something benevolent, according to the author of passage A. In the second paragraph of the passage, the author states that pecuniary (or monetary) aid is the easiest form of benevolence and that it requires the least amount of sacrifice. The remaining answer choices would all involve much greater amounts of sacrifice.

22. B: The phrase "mental exercise" is used in paragraph two of passage A to represent trying to see things from another person's point of view. In paragraph two, the author is discussing practicing benevolence by placing oneself in the situations of others. The author does not make mention of a person only taking their own perspective into consideration, improving an individual's memory recall, or strengthening a person's concentration.

23. C: In passage A, the term "heads" means manner. The author is discussing treating the practice of benevolent affections in almost the same way or manner as when considering the principles of justice. The remaining answer choices (discord, unlikeness, and opposition) do not make sense in this context.

24. D: The author of passage A states that justice can be exercised at the expense of benevolence, but this premise is not true the other way around ("benevolence is not to be exercised at the expense of justice"). Additionally, acts of benevolence require selflessness (not acts of justice). Finally, benevolence cannot take the place of the payment of debts that are owed per acts of justice.

25. A: The author of passage B says that giving food and money to beggars is seen as a weakness because that type of charity encourages laziness and corruption ("giving alms to common beggars" leads to an increase in "idleness and debauchery"). Although the remaining answer choices may indeed be true, they are not mentioned by the author of passage B.

26. B: The phrase "their gentle dominion over the breasts of men" refers to the social virtues, compassion and benevolence. In this paragraph, the author discusses how the goal of benevolence is to promote the happiness of society. The other answer choices are mentioned in paragraph five, but the author does not propose they rule over people's inner spirits ("the breasts of men").

27. C: The author's attitude toward luxury in passage B can best be described as objective. The author is impartial and simply represents facts and attitudes about luxury. The author does not say luxury is unworthy of consideration (dismissive), does not think romantically or sentimentally about luxury (nostalgic), and does not worship luxury (adoring).

28. D: Both passage A and passage B are written in first-person point of view. The author uses the first-person plural pronouns *we* and *us*. Second-person point of view would use the second-person pronoun, *you*. Third-person objective point of view would utilize third-person pronouns: *he, she, it,* or *they*, and the author would stay detached by telling the story. Finally, third-person limited omniscient view would also use the third-person pronouns, and the author would tell the entire story through the eyes of a single character.

29. A: The tone of passage A is objective, and the tone of passage B is earnest. The author is passage A is uninfluenced by emotion, while the author of passage B shows deeper feelings with the incorporation of words in all capital letters and the phrases and examples which he selected. A cynical tone is used when an author is mocking others, and a sarcastic tone is used when an author wishes to state a negative opinion in an ironic way. Finally, an excited tone is used when an author is very happy about events, and an ambivalent tone is used when an author is unsure or undecided about something.

30. B: Compared to the author of passage A, the author of passage B stresses the importance of exercising benevolence for the purposes of contributing to the overall happiness of society ("bestow happiness on human society"). The author of passage A mentions the importance of using benevolence to alleviate the distresses of others and to protect the reputations of others. The concept of attracting karma is not mentioned in either passage.

31. A: The main focus of the passage is a discussion of the causes, symptoms, and treatments associated with the various types of tenosynovitis. The remaining answer choices are all mentioned throughout the passage; however, they are separate, subordinate topics, not the main idea of the passage.

32. B: Placing the tendon at rest and possibly utilizing a splinting device is mentioned in the passage as a treatment for tenosynovitis crepitans of the ankle. Depositing urate of soda under the endothelial tendon covering is used to treat gouty tenosynovitis. Inducing hyperemia and making small incisions to release pus, as well as incorporating active and passive movements to prevent tendon adhesions, are both mentioned in the passage as treatments for suppurative tenosynovitis.

33. C: The statement is supported by the information found in the third paragraph, since the author states that only a small proportion of individuals who overuse their tendons are afflicted with the condition. This means that something else, such as the toxic factor that is mentioned, must be the reason.

34. D: According to the author, a patient having a sensation of crepitation or friction when the tendon is moved in its sheath would be presenting with a case of tenosynovitis crepitans. Visible nodular swellings are a clinical feature of gouty tenosynovitis. Finally, cellulitis and finger wound infections are both clinical features of suppurative tenosynovitis.

35. A: The authors use first-person point of view in this passage, which can be determined by the use of the first-person pronoun, *we*, in paragraph eight. Second-person point of view would involve the authors utilizing the second-person pronoun, *you*. Third-person limited omniscient view would involve the authors using third-person pronouns (*he, she, it,* or *they*), and the authors would tell the entire story through the eyes of a single character. Finally, third-person objective point of view would also involve the authors using the third-person pronouns: *he, she, it,* or *they*. However, the authors would stay detached and would not refer to themselves with the pronouns "I" or "we."

36. B: Pus spreading to the adjacent bones and joints of the wrist may be the end result of an infected wound of the finger, according to the passage. Based on the options provided, this answer choice would have the most finality for an individual suffering with this condition. The other answer choices are all complications that can result from an infected wound of the finger, but Choices *A* and *C* are not as serious as the infection spreading from the finger to the bones and joints of the wrist. Choice *C*, amputation of the finger, is explicitly mention has a condition that can *precede* suppurative tenosynovitis, so it is not a good answer choice for an end result.

37. C: The phrases are used in paragraph four to describe the friction associated with tendon movement in the sheath. The authors utilize the phrases to express how the friction can present in varying degrees. Milder friction can be "soft like the friction of snow"; more intense friction can "resemble the creaking of new leather." The other answer choices are simply additional features of tenosynovitis crepitans.

38. D: The authors state in paragraph six that tenosynovitis crepitans at the ankle can result from the pressure of ill-fitting boots and that the individual affected would experience pain when raising his or her heels from the ground. Therefore, based on this passage, if a construction worker wearing tight boots experiences pain when raising his heel, he will more than likely be diagnosed with tenosynovitis crepitans.

39. A: The authors included the words "washerwomen" and "riveters" in paragraph five to give examples of workers whose repetitive motions on the job can result in tenosynovitis crepitans. The authors make no mention in the passage of the importance of healthcare for blue-collar workers or ergonomics in the workplace. Finally, the authors mention that recurrent attacks of inflammation are likely to occur in individuals suffering from gouty tenosynovitis. However, they do not provide any examples related to this in the passage.

40. B: Based on this passage, if a woman presents to her doctor with a nodular swelling on her thumb the size of a blueberry, she will more than likely be diagnosed with gouty tenosynovitis. This is the only affliction the authors mention is characterized by the formation of visible nodular swellings that vary in size from a pea to a cherry.

ACT Science

Interpretation of Data

Tables, charts, and graphs can be used to visually organize, categorize, and compare data. There are a variety of types, with carrying its own benefits. Answering a question relating to a table, chart, or graph, begins by reading the question and any accompanying information or introduction thoroughly to determine what is being asked and what is unknown. Then, the title of the table, chart, or graph, axes labels and units, and a key, if present, should be read. These provide information about how the data is organized and how it should be interpreted.

Tables organize data into vertical columns and horizontal rows. They offer the ability to neatly present a lot of data in a small space. For example, a table may be used to show the growth rate of different plants each week for six months during an experiment where the amount of fertilizer was varied to assess its impact on plant growth. By interpreting the table, one may observe which fertilizer condition resulted in the most growth, which was associated with least growth, and over which weeks the growth was greatest. In using a bar graph to display average monthly gasoline prices in different states, prices can be compared between states at different times of the year. Graphs are also a useful way to show change in variables over time, as in a line graph, or percentages of a whole, as in a pie graph.

The table below relates the number of items to the total cost. The table shows that 1 item costs $5. By looking at the table further, 5 items cost $25, 10 items cost $50, and 50 items cost $250. This cost can be extended for any number of items. Since 1 item costs $5, then 2 items would cost $10. Though this information isn't in the table, the given price can be used to calculate unknown information.

Number of Items	1	5	10	50
Cost ($)	5	25	50	250

A bar graph displays data via bars of different heights. It is useful when comparing two or more items or when seeing how a quantity changes over time. It has both a horizontal and vertical axis, which help communicate what variables are measured and in what units; thus, interpreting bar graphs involves recognizing what each bar represents and connecting that to the two variables. The bar graph below shows the number of birds observed feeding on six different types of seeds at bird feeders over two hours. The height of the bar indicates the number of birds who consumed that particular type of food. For example, five birds ate corn, and two had thistle seeds.

By comparing the bars, it's obvious that more birds ate the mixed seeds rather than sorghum during this experiment.

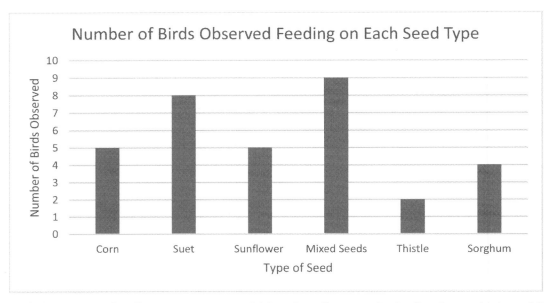

A line graph is a way to visually compare two variables. On a **line graph**, the line for each plotted line can be followed to notice the change in the variables. This change can be seen in how the line rises or falls, known as its slope, or rate of change. In other words, the trend of the data can be easily observed. While not always the case, the *x*-axis on line graphs often represents the variable time in units such as days, hours, months, etc. However, readers of line graphs should always assess what the axes represent and the units used. If there are multiple lines, a comparison can be made between what the two lines represent. For example, the following line graph shows monthly rainfall in two cities over the 2017 year. The lighter line represents the rainfall in Atlanta, and the darker line represents the rainfall in Boston. Looking at the lighter line alone, the rainfall increases through March, then decreases through May, increases again for June, etc. Boston has a somewhat different trend, shown by the darker line. The difference in rainfall between the two cities can also be calculated by finding the difference between the lighter line and the darker line; it is easy to see that Atlanta, on average, received more rain than Boston. The range in the amount of monthly rainfall for each city can also be calculated by finding the highest and lowest point for each line and finding the difference. For example, for Boston, the months with the most rain were May and June, each with 5 inches, while the lowest months were June and December, each with 1 inch. Thus, the range in monthly rainfall was 4 inches (5 – 1 in.).

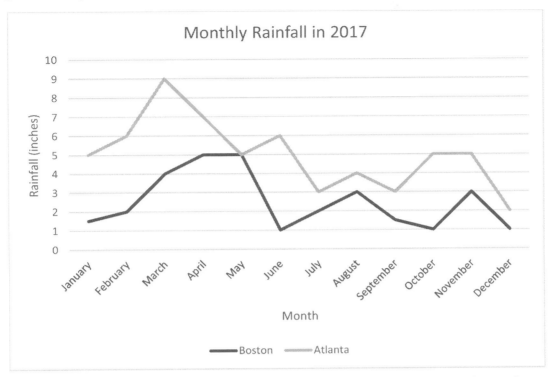

Pie charts use a circular representation of data to highlight the numerical proportion that each category comprises of the whole. The arc length of each pie slice is proportional to the amount that category of the pie individually represents of the whole. Pie charts allow for making quick comparisons about the relative sizes of groups or categories that together comprise a larger group or set. The following pie chart is a simple example showing the types of rocks found on a hiking trip. The labels show that black represents igneous rocks, light gray represents sedimentary rocks, and dark gray represents metamorphic rocks. As the whole pie represents 100%, each slice represents a certain percentage of the whole. For example, of the total number of rocks found, 16% were igneous. It is also easy to see that the majority of rocks found were sedimentary.

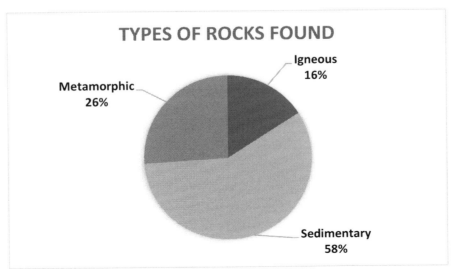

The fact that a circle is 360 degrees is used to create a pie chart. Because each piece of the pie is a percentage of a whole, that percentage is multiplied times 360 to get the number of degrees each piece represents. In the example above, each piece is $\frac{1}{3}$ of the whole, so each piece is equivalent to 120 degrees. Together, all three pieces add up to 360 degrees.

Stacked bar graphs, also used fairly frequently, are used when comparing multiple variables at one time. They combine the organization of bar graphs with the proportionality aspect of pie charts. The following is an example of a stacked bar graph that shows the caloric content of four different breakfast options. Each bar graph is broken up further into the contribution of calories from each of the three major macronutrients.

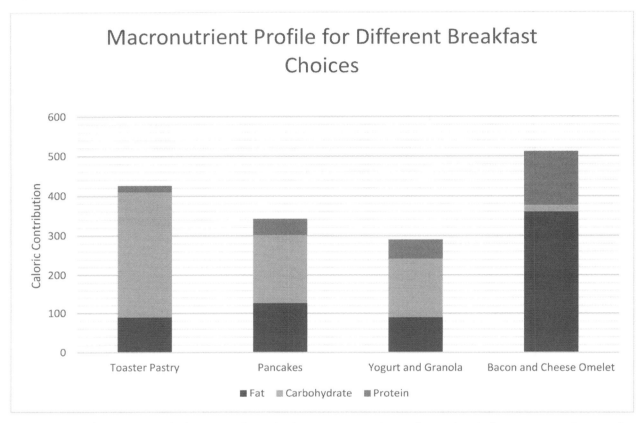

To determine how many calories come from fat in toaster pastries, refer to the darkest gray portion on the bottom of the Toaster Pastry bar, resulting in 90 calories.

A scatterplot is another way to represent paired data. Like a line graph, it uses Cartesian coordinates, which means that it has both a horizontal and vertical axis. Each data point is represented by a dot on the graph, but unlike a line graph, these dots are not connected with a line. For example, the following is a scatterplot showing people's attention span versus age.

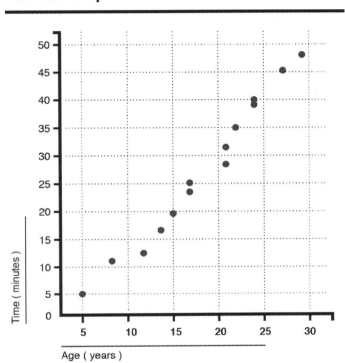

Attention Span

It can be seen that there is a trend or correlation in the data set. As age increases, so too does attention span. If the data resembles a straight line, as it does in this case, the data is has a positive linear association.

Descriptive Statistics

Mean, Median, Mode, and Range
One way information can be interpreted from tables, charts, and graphs is through descriptive statistics. The three most common calculations for a set of data are the measures of central tendency: mean, median, and mode. Measures of central tendency are helpful in comparing two or more different sets of data and making generalizations about data collected. The **mean** refers to the average and is equal to the sum of all values divided by the total number of data entries. For example, if you ran 35 miles total over in 6 bowling games, your mean distance per run was $\frac{35}{6} = 5.83$ miles. Students frequently use mean, especially when calculating what grade they need to receive on a final exam to earn a desired grade in a class.

The **median** is found by ordering values from least to greatest and selecting the middle value. If there's an even number of values, then the mean of the two middle amounts must be calculated to obtain the median. For example, the median of the set of weights of students in pounds 95, 107, 112, 135, and 145 pounds is 112 pounds. The median set of weights of students in pounds 95, 107, 112, 135, 145, and 148 is 140 pounds, which is the mean of 135 and 145 pounds.

The **mode** is the value that occurs the most. The mode of the data set {1, 6, 2, 5, 5, 8, 2} actually refers to two numbers: 1 and 5. In this case, the data set is bimodal because it has two modes. A data set can have no mode if no value is repeated. Another useful statistic is range. The **range** for a set of data refers to the difference between the highest and lowest value.

Standard Deviation, Quartiles, and Percentiles
A set of data can be described using its **standard deviation**, or how spread apart the values in a data set are. A lower standard deviation indicates that the data points in the set do not differ significantly from the mean. Standard deviation is calculated as the square root of the variance, so the formula for the standard deviation of a data set is:

$$s = \sqrt{\frac{\Sigma(x - \bar{x})^2}{n - 1}}$$

where x is each value in the dataset, $\bar{x}$ is the mean, and n is the total number of data points in the set. Recall that variance is the average of the squared differences of each data point from the mean.

A dataset can be broken up into four equal parts. The three **quartiles** Q_1, Q_2, and Q_3 split up the data into four equal parts. Q_1 is the first quartile, and one-quarter of the data falls on or below it. Q_2 is the second quartile, also the median, and one-half of the data falls on or below it. Q_3 is the third quartile, and three-quarters of the data falls on or below it. The **interquartile range (IQR)** of a dataset gives the range of the middle 50 percent of the data, and its formula is:

$$IQR = Q_3 - Q_1$$

Similar to quartiles, **deciles** divide a dataset into ten equal parts, and **percentiles** divide a dataset into one hundred equal parts. For example, the 90th percentile refers to splitting up a dataset into the bottom 90 percent of the data and the top 10 percent of the data.

Elementary Probability

A **probability experiment** is an action that causes specific results, such as counts or measurements. The result of such an experiment is known as an **outcome,** and the set of all potential outcomes is known as the **sample space.** An **event** consists of one or more of those outcomes. For example, consider the probability experiment of tossing a coin and rolling a six-sided die. The coin has two possible outcomes—a heads or a tails—and the die has six possible outcomes—rolling each number 1–6. Therefore, the sample space has twelve possible outcomes: a heads or a tails paired with each roll of the die.

A **simple event** is an event that consists of a single outcome. For instance, selecting a queen of hearts from a standard fifty-two-card deck is a simple event; however, selecting a queen is not a simple event because there are four possibilities.

Classical, or **theoretical, probability** is when each outcome in a sample space has the same chance to occur. The probability for an event is equal to the number of outcomes in that event divided by the total number of outcomes in the sample space. For example, consider rolling a six-sided die. The probability of rolling a 2 is $\frac{1}{6}$, and the probability of rolling an even number is $\frac{3}{6}$, or $\frac{1}{2}$, because there are three even numbers on the die. This type of probability is based on what should happen in theory but not what actually happens in real life.

Empirical probability is based on actual experiments or observations. For instance, if a die is rolled eight times, and a 1 is rolled two times, the empirical probability of rolling a 1 is $\frac{2}{8} = \frac{1}{4}$, which is higher than the theoretical probability. The **Law of Large Numbers** states that as an experiment is completed repeatedly, the empirical probability of an event should get closer to the theoretical probability of an event.

Probabilities range from 0 to 1. The closer the probability of an event occurring is to 0, the less likely it will occur. The closer it is to 1, the more likely it is to occur.

The **addition rule** is necessary to find the probability of event A or event B occurring or both occurring at the same time. If events A and B are **mutually exclusive** or **disjoint,** which means they cannot occur at the same time:

$$P(A \text{ or } B) = P(A) + P(B)$$

If events A and B are not mutually exclusive:

$$P(A \text{ or } B) = P(A) + P(B) - P(A \text{ and } B)$$

where $P(A \text{ and } B)$ represents the probability of event A and B both occurring at the same time. An example of two events that are mutually exclusive are rolling a 6 on a die and rolling an odd number on a die. The probability of rolling a 6 or rolling an odd number is:

$$\frac{1}{6} + \frac{3}{6} = \frac{4}{6} = \frac{2}{3}$$

Rolling a 6 and rolling an even number are not mutually exclusive because there is some overlap. The probability of rolling a 6 or rolling an even number is:

$$\frac{1}{6} + \frac{3}{6} - \frac{1}{6} = \frac{3}{6} = \frac{1}{2}$$

Conditional Probability

The **multiplication rule** is necessary when finding the probability that an event A occurs in a first trial and event B occurs in a second trial, which is written as $P(A \text{ and } B)$. This rule differs if the events are independent or dependent. Two events A and B are **independent** if the occurrence of one event does not affect the probability that the other will occur. If A and B are not independent, they are **dependent,** and the outcome of the first event somehow affects the outcome of the second. If events A and B are independent:

$$P(A \text{ and } B) = P(A)P(B)$$

and if events A and B are dependent:

$$P(A \text{ and } B) = P(A)P(B|A)$$

where $P(B|A)$ represents the probability event B occurs given that event A has already occurred.

$P(B|A)$ represents **conditional probability**, or the probability of event B occurring given that event A has already occurred. $P(B|A)$ can be found by dividing the probability of events A and B both occurring by the probability of event A occurring using the formula:

$$P(B|A) = \frac{P\,(A \text{ and } B)}{P(A)}$$

and represents the total number of outcomes remaining for B to occur after A occurs. This formula is derived from the multiplication rule with dependent events by dividing both sides by $P(A)$. Note that $P(B|A)$ and $P(A|B)$ are not the same. The first quantity shows that event B has occurred after event A, and the second quantity shows that event A has occurred after event B. To incorrectly interchange these ideas is known as **confusion of the inverse.**

Consider the case of drawing two cards from a deck of fifty-two cards. The probability of pulling two queens would vary based on whether the initial card was placed back in the deck for the second pull. If the card is placed back in, the probability of pulling two queens is:

$$\frac{4}{52} \times \frac{4}{52} = 0.00592$$

If the card is not placed back in, the probability of pulling two queens is:

$$\frac{4}{52} \times \frac{3}{51} = 0.00452$$

When the card is not placed back in, both the numerator and denominator of the second probability decrease by 1. This is due to the fact that, theoretically, there is one fewer queen cards in the deck, and there is one fewer cards overall in the deck as well.

Probability Distributions

A **discrete random variable** is a set of values that is either finite or countably infinite. If there are infinitely many values, being **countable** means that each individual value can be paired with a natural number. For example, the number of coin tosses before getting heads could potentially be infinite, but the total number of tosses is countable. Each toss refers to a number, like the first toss, second toss, etc. A **continuous random variable** has infinitely many values that are not countable. The individual items cannot be enumerated; an example of such a set is any type of measurement. There are infinitely many heights of human beings due to decimals that exist within each inch, centimeter, millimeter, etc. Each type of variable has its own **probability distribution**, which calculates the probability for each potential value of the random variable. Probability distributions exist in tables, formulas, or graphs. The expected value of a random variable represents what the mean value should be in either a large sample size or after many trials. According to the **Law of Large Numbers**, after many trials, the actual mean and that of the probability distribution should be very close to the expected value. The **expected value** is a weighted average that is calculated as:

$$E(X) = \sum x_i p_i$$

where x_i represents the value of each outcome, and p_i represents the probability of each outcome. The expected value if all of the probabilities are equal is:

$$E(X) = \frac{x_1 + x_2 + \cdots + x_n}{n}$$

Expected value is often called the **mean of the random variable** and is known as a **measure of central tendency** like mean and mode.

A **binomial probability distribution** is a probability distribution that adheres to some important criteria. The distribution must consist of a fixed number of trials where all trials are independent, each trial has an outcome classified as either success or failure, and the probability of a success is the same in each trial. Within any binomial experiment, x is the number of resulting successes, n is the number of trials, P is the probability of success within each trial, and $Q = 1 - P$ is the probability of failure within each trial. The probability of obtaining x successes within n trials is:

$$\binom{n}{x} P^x (1 - P)^{n-x}, \text{ where } \binom{n}{x} = \frac{n!}{x!(n-x)!}$$

is called the **binomial coefficient.** A binomial probability distribution could be used to find the probability of obtaining exactly two heads on five tosses of a coin. In the formula, $x = 2$, $n = 5$, $P = 0.5$, and $Q = 0.5$.

A **uniform probability distribution** exists when there is constant probability. Each random variable has equal probability, and its graph is a rectangle because the height, representing the probability, is constant.

Finally, a **normal probability distribution** has a graph that is symmetric and bell-shaped; an example using body weight is shown here:

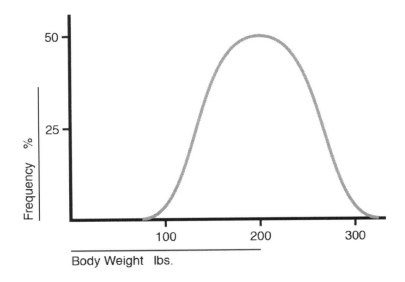

Population percentages can be estimated using normal distributions. For example, the probability that a data point will be less than the mean is 50 percent. The **Empirical Rule** states that 68 percent of the data falls within 1 standard deviation of the mean, 95 percent falls within 2 standard deviations of the mean, and 99.7 percent falls within 3 standard deviations of the mean. A **standard normal distribution** is a normal distribution with a mean equal to 0 and standard deviation equal to 1. The area under the entire curve of a standard normal distribution is equal to 1.

Counting Methods

The total number of events in the sample space must be known to solve probability problems. Different methods can be used to count the number of possible outcomes, depending on whether different arrangements of the same items are counted only once or separately. **Permutations** are arrangements in which different sequences are counted separately. Therefore, order matters in permutations. **Combinations** are arrangements in which different sequences are not counted separately. Therefore, order does not matter in combinations. For example, if 123 is considered different from 321, permutations would be discussed. However, if 123 is considered the same as 321, combinations would be considered.

If the sample space contains n different permutations of n different items and all of them must be selected, there are $n!$ different possibilities. For example, five different books can be rearranged:

$$5! = 5 \times 4 \times 3 \times 2 \times 1 = 120 \text{ times}$$

The probability of one person randomly ordering those five books in the same way as another person is $1/120$. A different calculation is necessary if a number less than n is to be selected or if order does not matter. In general, the notation $P(n,r)$ represents the number of ways to arrange r objects from a set of n if order does matter, and:

$$P(n,r) = \frac{n!}{(n-r)!}$$

Therefore, in order to calculate the number of ways five books can be arranged in three slots if order matters, plug $n = 5$ and $r = 3$ in the formula to obtain:

$$P(5,3) = \frac{5!}{(5-3)!} = \frac{5!}{2!} = 60$$

Secondly, $C(n,r)$ represents the total number of r combinations selected out of n items when order does not matter, and:

$$C(n,r) = \frac{n!}{(n-r)! \, r!}$$

Therefore, the number of ways five books can be arranged in three slots if order does not matter is:

$$C(5,3) = \frac{5!}{(5-3)! \, 3!} = 10.$$

The following relationship exists between permutations and combinations:

$$C(n,r) = \frac{P(n,r)}{r!}$$

Scientific Investigation

Scientific investigation is seeking an answer to a specific question using the scientific method. The **scientific method** is a procedure that is based on an observation. For example, if someone observes something strange, and they want to understand the process, they might ask how that process works. If we were to use the scientific method to understand the process, we would formulate a hypothesis. A **hypothesis** explains what you expect to happen from the experiment. Then, an experiment is conducted, data from the experiment is analyzed, and a conclusion is reached. Below are steps of the scientific method in a practical situation:

- Observation: A student observes that every time he eats a big lunch, he falls asleep in the next period.

- Experimental Question: The question the student asks is, what is the relationship between eating large meals and energy levels?

- Formulate a Hypothesis: The student hypothesizes that eating a big meal will lead to lower energy due to the body using its energy to digest the food.

- Conduct an Experiment (Materials and Procedure): The student has his whole class eat a lunch twice the size of their normal one for a week, then he measures their energy levels in the next period. The next week, the student has them eat a small lunch, and then measures their energy levels in the next period.

- Analyze Data: The data in the student's experiment showed that in the first week, sixty percent of the students felt sleepy in the next period class and that in the second week, only twenty percent of the students felt sleepy in the next period class.

- Conclusion: The student concluded that the data supported his hypothesis: student energy levels decrease after the consumption of a large meal. However, the student recognized that variables would need to be controlled further to make more definitive conclusions.

An important component of scientific research after the experiment is conducted is called **peer review**. This is when the article written about the experiment is sent to other scientists to "review." The reviewers offer feedback so that the editors can decide whether or not to publish the findings in a scholarly or scientific journal.

Evaluation of Models, Inferences, and Experimental Results

Identifying Basic Scientific Measurement Using Laboratory Tools

Measuring Force

A Newton meter or force meter is a standardized instrument for measuring the force exerted by different elements in the universe that cause movement when they act upon objects by pulling, pushing, rotating, deforming, or accelerating them, such as gravity, friction, or tension. Force is measured in units called Newtons (N), named after Sir Isaac Newton, who defined the laws of motion, including forces. One common example of a force meter is a bathroom scale. When someone steps on the scale, the force exerted on the platform is measured in the form of units of weight. Force meters contain springs, rubber bands, or other elastic materials that stretch in proportion to force applied. A Newton meter displays 50-

Newton increments, with four 10-Newton marks in between. Thus, if a Newton meter's needle rests on the third mark between 100 and 150, it indicates a measurement of 130N.

Measuring Temperature

As those familiar with word roots can determine, **thermometers** measure heat or temperature (*thermo-* is from Greek for heat and *meter* from Greek for measure). Some thermometers measure outdoor or indoor air temperature, and others measure body temperature in humans or animals. Meat thermometers measure temperature in the center of cooked meat to ensure sufficient cooking to kill bacteria. Refrigerator and freezer thermometers measure internal temperatures to ensure sufficient coldness. Although digital thermometers eliminate the task of reading a thermometer scale by displaying a specific numerical reading, many thermometers still use visual scales. In these, increments are typically two-tenths of a degree. However, basal thermometers, which are more sensitive and accurate, and are frequently used to track female ovulation cycles for measuring and planning fertility, display increments of one-tenth of a degree. Whole degrees are marked by their numbers; tenths of degrees are the smallest marks in between.

Measuring Length

Rulers, yardsticks, tape measures, and other standardized instruments are for measuring short distances. Standard rulers and similar measures show distances in increments of feet, inches, and halves, quarters, eights, and sixteenths of an inch. Tape measures have markings indicating one yard or three feet, six feet, and other common multiple-feet measurements. Each inch on a ruler is marked by number from 1 to 12 (or 1 to 36 on a yardstick). Half-inches are the longest of the non-numbered line marks, then quarter-inches, then eighth-inches, then sixteenth-inches. Some American-made rulers also include metric measurements.

Using the Vernier Scale

French mathematician and inventor Pierre Vernier originated the **Vernier measurement principle** and corresponding scale in 1631. When measuring small increments using other instruments, it can be impossible to determine precisely small fractions within already small divisions. Vernier's invention solved this problem with simple elegance. For example, dividing millimeters in the metric system or 16ths of an inch in the English system into 10ths or smaller portions is impractical, inaccurate, and unreadable without using magnifiers. A metric **Vernier caliper** features a main scale and a Vernier scale, i.e. a parallel, sliding scale with ten or eleven marks, spaced equally within themselves but differently from the main scale spacing. English Vernier calipers often have $\frac{1}{16}$" main scale divisions and 8 divisions on the Vernier scale, so they can measure lengths as small as $\frac{1}{28}$". To read a Vernier caliper, close its jaws on the object being measured, note the main scale number, rounding down to the next smaller marking, and see which Vernier scale mark aligns with it to obtain the additional fraction.

Using a Micrometer Caliper

A **micrometer caliper** is similar to a Vernier caliper in that both have a main scale plus a movable scale for measuring fractions of small numbers. Like many medical instruments, micrometer calipers are typically metrical, measuring in millimeters. The main scale markings are in $\frac{1}{2}$ millimeters. A uniform, precise screw moves the micrometer caliper's jaw. It has a rotating thimble on its handle, marked in 50 equal divisions. Rotating the thimble once moves the screw $\frac{1}{2}$ millimeter along the main scale, enabling the user to read measurements as small as one-hundredth of a millimeter. Most micrometers display $\frac{1}{2}$ mm markings on the thimble on the opposite side from the main scale markings on the micrometer sleeve, making it easier to read. Users require some practice, e.g. rounding up $\frac{1}{2}$ mm if a reading is in the top half of a millimeter; only tightening the caliper jaws using the slip clutch to close them snugly without bending them; and making a zero correction using a special wrench so fully closing the jaws yields a zero reading if it did not already.

Critiquing a Scientific Explanation Using Logic and Evidence

According to the scientific method, one must either observe something as it is happening or as a constant state, or prove it through experimenting, to say it is a fact. Scientific experiments are found reliable when they can be replicated and produce comparable results regardless of who is conducting them or making the observations. However, historically many beliefs have been considered facts at the time, only to be disproven later when objective evidence became accessible. For instance, in ancient times many people believed the Earth was flat based on limited visual evidence. Ancient Greek philosophers Pythagoras in the 6th century BCE and Aristotle in the 4th century BCE believed the earth was spherical. This was eventually borne out by observational evidence. Aristotle estimated the Earth's circumference; Eratosthenes measured it c. 240 BCE; in the 2nd century CE, Ptolemy had mapped the globe and developed the latitude, longitude, and climes system.

While many phenomena have been established by science as facts, technically, many others are actually theories, though many people mistakenly consider them facts. As one example, gravity per se is a fact; however, the scientific explanation of how gravity functions is a theory. Although a theory is not the same as a fact, theories are not merely speculative ideas. A scientific theory must be tested thoroughly and then applied to established facts, hypotheses, and observations. Moreover, for a theory to be accepted or even considered scientifically-sound, it must relate and explain a broad scope of observations that would not be related without the theory. People sometimes state opinions as if they were facts to persuade others. For example, an advertisement might claim a product is the best without concrete supporting evidence. To evaluate information, one must consider both its source and any supporting evidence to ascertain its veracity.

Relationships Among Events, Objects, and Procedures

When we determine relationships among events, objects, and procedures, we are better able to understand the world around us and make predictions based on that understanding. With regards to relationships among events and procedures, we will look at cause and effect. For relationships among objects, we will take a look at Newton's Laws.

Cause
The **cause** of a particular event is the thing that brings it about. A causal relationship may be partly or wholly responsible for its effect, but sometimes it's difficult to tell whether one event is the sole cause of another event. For example, lung cancer can be caused by smoking cigarettes. However, sometimes, lung cancer develops even in a nonsmoker, and that tells us that there may be other factors involved in lung cancer besides smoking. It's also easy to make mistakes when considering causation. One common mistake is mistaking correlation for causation. For example, say that in the year 2008 a volcano erupted in California, and in that same year, the number of infant deaths increased by ten percent. If we automatically assume, without looking at the data, that the erupting volcano *caused* the infant deaths, then we are mistaking correlation for causation. The two events might have happened at the same time, but that does not necessarily mean that one event caused the other. Relationships between events are never absolute; there are a myriad of factors that can be traced back to their foundations, so we must be thorough with our evidence in proving causation.

Effect

An **effect** is the result of something that occurs. For example, the Nelsons have a family dog. Every time the dog hears thunder, the dog whines. In this scenario, the thunder is the cause, and the dog's whining is the effect. Sometimes, a cause will produce multiple effects. Let's say we are doing an experiment to see what the effects of exercise are in a group of people who are not usually active. After about four weeks, the group experienced weight loss, increased confidence, and higher energy. We start out with a cause: exercising more. From that cause, we have three known effects that occurred within the group: weight loss, increased confidence, and higher energy. Cause and effect are important terms to understand when conducting scientific experiments.

Newton's Laws

Newton's laws of motion describe the relationship between an object and the forces acting upon that object, and its movement responding to those forces. Newton has three laws:

1. Law of Inertia: An object at rest stays at rest and an object in motion stays in motion unless otherwise acted upon by an outside force. For example, gravity is an outside force that will affect the speed and direction of a ball; when we throw a ball, the ball will eventually decrease in speed and fall to the ground because of the outside force of gravity. However, if the ball were kicked in space where there is no gravity, the ball would go the same speed and direction forever unless it hits another object in space or falls into another gravity field.

2. Newton's Second Law: This law states that the heavier an object, the more force required to move it. This force has to do with acceleration. For example, if you use the same amount of force to push both a golf cart and an eighteen-wheeler, the golf cart will have more acceleration than the truck because the eighteen-wheeler has more mass than the golf cart.

3. Newton's Third Law: This law states that for every action, there is an opposite and equal reaction. For example, if you are jumping on a trampoline, you are experiencing Newton's third law of motion. When you jump down on the trampoline, the opposite reaction pushes you up into the air.

Biology

Biology Basics

Atoms are the smallest units of all matter and make up all chemical elements. They each have three parts: protons, neutrons, and electrons. **Protons** are found in the nucleus of an atom and have a positive electric charge. They have a mass of about one atomic mass unit. The number of protons in an element is referred to as the element's **atomic number**. Each element has a unique number of protons, and therefore a unique atomic number. **Neutrons** are also found in the nucleus of atoms. These subatomic particles have a neutral charge, meaning that they do not have a positive or negative electric charge. Their mass is slightly larger than that of a proton. Together with protons, they are referred to as the **nucleons** of an atom. The **atomic mass** number of an atom is equal to the sum of the protons and neutrons in the atom. **Electrons** have a negative charge and are the smallest of the subatomic particles. They are located outside the nucleus in **orbitals,** which are shells that surround the nucleus. If an atom has an overall neutral charge, it has an equal number of electrons and protons. If it has more protons than electrons or vice versa, it becomes an **ion.** When there are more protons than electrons, the atom is a positively-charged ion, or **cation.** When there are more electrons than protons, the atom is a negatively-charged ion, or **anion.**

The location of electrons within an atom is more complicated than the locations of protons and neutrons. Within the orbitals, electrons are always moving. They can spin very fast and move upward, downward, and sideways. There are many different levels of orbitals around the atomic nucleus, and each orbital has a different capacity for electrons. The electrons in the orbitals closest to the nucleus are more tightly bound to the nucleus of the atom. There are three main characteristics that describe each orbital. The first is the **principle quantum number**, which describes the size of the orbital. The second is the **angular momentum quantum number**, which describes the shape of the orbital. The third is the **magnetic quantum number,** which describes the orientation of the orbital in space.

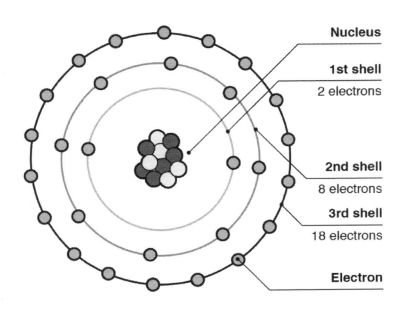

Another important characteristic of electrons is their ability to form covalent bonds with other atoms to form molecules. A **covalent bond** is a chemical bond that forms when two atoms share the same pair or pairs of electrons. There is a stable balance of attraction and repulsion between the two atoms. There are several types of covalent bonds. **Sigma bonds** are the strongest type of covalent bond and involve the head-on overlapping of electron orbitals from two different atoms. **Pi bonds** are a little weaker and involve the lateral overlapping of certain orbitals. While single bonds between atoms, such as between carbon and hydrogen, are generally sigma bonds, double bonds, such as when carbon is double-bonded to an oxygen atom, are usually formed from one sigma bond and one pi bond.

The Cell

All living organisms are made up cells. **Cells** are considered the basic functional unit of organisms and the smallest unit of matter that is living. Most organisms are multicellular, which means that they are made up of more than one cell and often they are made up of a variety of different types of cells. Cells contain **organelles**, which are the little working parts of the cell, responsible for specific functions that keep the cell and organism alive.

Plant and animal cells have many of the same organelles but also have some unique traits that distinguish them from each other. Plants contain a cell wall, while animal cells are only surrounded by a phospholipid plasma membrane. The cell wall is made up strong, fibrous polysaccharides and proteins. It protects the cell from mechanical damage and maintains the cell's shape. Inside the cell wall, plant cells also have plasma membrane. The plasma membrane of both plant and animal cells is made up of two layers of phospholipids, which have a hydrophilic head and hydrophobic tails. The tails converge towards each other on the inside of the bilayer, while the heads face the interior of the cell and the exterior environment. **Microvilli** are protrusions of the cell membrane that are only found in animal cells. They increase the surface area and aid in absorption, secretion, and cellular adhesion. Chloroplasts are also only found in plant cells. They are responsible for photosynthesis, which is how plants convert sunlight into chemical energy.

The list below describes major organelles that are found in both plant and animal cells:

- Nucleus: The nucleus contains the DNA of the cell, which has all of the cells' hereditary information passed down from parent cells. DNA and protein are wrapped together into chromatin within the nucleus. The nucleus is surrounded by a double membrane called the nuclear envelope.

- Endoplasmic Reticulum (ER): The ER is a network of tubules and membranous sacs that are responsible for the metabolic and synthetic activities of the cell, including synthesis of membranes. Rough ER has ribosomes attached to it, while smooth ER does not.

- Mitochondrion: The mitochondrion is essential for maintaining regular cell function and is known as the powerhouse of the cell. It is where cellular respiration occurs and where most of the cell's ATP is generated.

- Golgi Apparatus: The Golgi Apparatus is where cell products are synthesized, modified, sorted, and secreted out of the cell.

- Ribosomes: Ribosomes make up a complex that produces proteins within the cell. They can be free in the cytosol or bound to the ER.

Cellular Respiration

Cellular respiration in multicellular organisms occurs in the mitochondria. It is a set of reactions that converts energy from nutrients to ATP and can either use oxygen in the process, which is called **aerobic respiration**, or not, which is called **anaerobic respiration**.

Aerobic respiration has two main parts, which are the citric acid cycle, also known as Krebs cycle, and oxidative phosphorylation. Glucose is a commonly-found molecule that is used for energy production within the cell. Before the citric acid cycle can begin, the process of glycolysis converts glucose into two pyruvate molecules. Pyruvate enters the mitochondrion, is oxidized, and then is converted to a compound called acetyl CoA. There are eight steps in the citric acid cycle that start with acetyl CoA and convert it to oxaloacetate and NADH. The oxaloacetate continues in the citric acid cycle and the NADH molecule moves on to the oxidative phosphorylation part of cellular respiration. Oxidative phosphorylation has two main steps, which are the electron transport chain and chemiosmosis. The mitochondrial membrane has four protein complexes within it that help to transport electrons through the inner mitochondrial matrix.

Electrons and protons are removed from NADH and $FADH_2$ and then transported along these and other membrane complexes. Protons are pumped across the inner membrane, which creates a gradient to draw electrons to the intermembrane complexes. Two mobile electron carriers, ubiquinone and cytochrome C, are also located in the inner mitochondrial membrane. At the end of these electron transport chains, the electrons are accepted by O_2 molecules and water is formed with the addition of two hydrogen atoms. Chemiosmosis occurs in an ATP synthase complex that is located next to the four electron transport complexes. As the complex pumps protons from the intermembrane space to the mitochondrial matrix,

ADP molecules become phosphorylated and ATP molecules are generated. Approximately four to six ATP molecules are generated during glycolysis and the citric acid cycle and twenty-six to twenty-eight ATP molecules are generated during oxidative phosphorylation, which makes the total number of ATP molecules generated during aerobic cellular respiration approximately thirty to thirty-two.

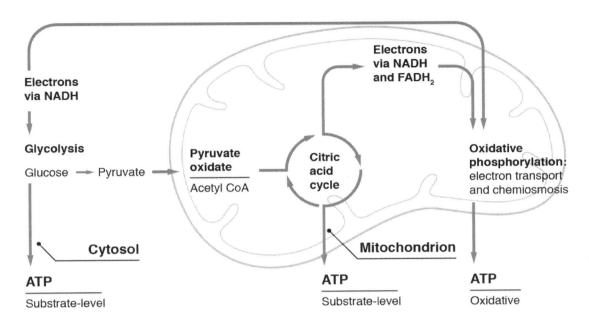

Since not all environments are oxygen-rich, some organisms must find alternate ways to extract energy from nutrients. The process of anaerobic respiration is similar to that of aerobic respiration in that protons and electrons are removed from nutrient molecules and are passed down an electron transport chain, with the end result being ATP synthesis. However, instead of the electrons being accepted by oxygen molecules at the end of the electron transport chain, they are accepted by either sulfate or nitrate molecules. Anaerobic respiration is mostly used by unicellular organisms, or prokaryotic organisms.

Photosynthesis

Photosynthesis is a set of reactions that occur to convert light energy into chemical energy. The chemical energy is then stored as sugar and other organic molecules inside the organism or plant. Within plants, the photosynthetic process takes place within chloroplasts. The two stages of photosynthesis are the light reactions and the Calvin cycle. Within chloroplasts, there are membranous sacs called **thylakoids** and within the thylakoids is the green pigment called **chlorophyll**. The light reactions take place in the chlorophyll. The Calvin cycle takes place in the **stroma,** or inner space, or the chloroplasts.

During the light reactions, light energy is absorbed by chlorophyll. First, a light-harvesting complex, called photosystem II (PS II), absorbs photons from light that enters the chlorophyll and then passes it onto a reaction-center complex. Once the photon enters the reaction-center complex, it causes a special pair of chlorophyll *a* molecules to release an electron. The electron is accepted by a primary electron acceptor molecule, while at the same time, a water molecule is dissociated into two hydrogen atoms, one oxygen atom, and two electrons. These electrons are transferred to the chlorophyll *a* molecules that just lost their electrons. The electrons that were released from the chlorophyll *a* molecules move down an electron transport chain using an electron carrier, called plastoquinone, a cytochrome complex, and a protein, called plastocyanin. At the end of the chain, the electrons reach another light-harvesting complex, called photosystem I (PS I). While in the cytochrome complex, the electrons cause protons to be pumped into the thylakoid space, which in turn provides energy for ATP molecules to be produced. A primary electron acceptor molecule accepts the electrons that are released from PS I and then passes them onto another electron transport chain, which includes the protein ferredoxin. At the end of the light reactions, electrons are transferred from ferredoxin to NADP+, producing NADPH. The ATP and NADPH that are produced through the light reactions are used as energy to drive the Calvin cycle forward.

The three phases of the Calvin cycle are carbon fixation, reduction, and regeneration of the CO_2 acceptor. Carbon fixation occurs when CO_2 is introduced into the cycle and attaches to a five-carbon sugar, called ribulose bisphosphate (RuBP). A six-carbon sugar is split into two three-carbon sugar molecules, known as 3-phosphoglycerate. Next, during the reduction phase, an ATP molecule loses a phosphate group and becomes ADP. The phosphate group attaches to the 3-phosphoglycerate molecule, making it 1,3-bisphosphate. Then, an NADPH molecule donates two electrons to this new molecule, causing it to lose a phosphate group and become glyceraldehyde 3-phosphate (G3P), a sugar molecule. At the end of the cycle, one G3P molecule exits the cycle and is used by the plant for energy. Five other G3P molecules continue in the cycle to regenerate RuBP molecules, which are the CO_2 acceptors of the cycle. When every photon has been used up, three RuBP molecules are formed from the rearrangement of five G3P molecules and wait for the cycle to start again. It takes three turns of the cycle and three CO_2 molecules entering the cycle to generate just one G3P molecule.

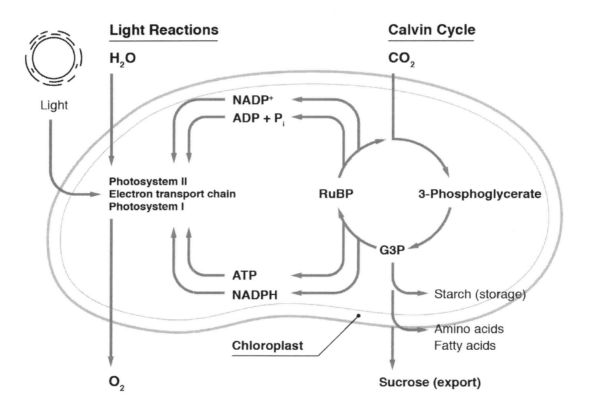

Cellular Reproduction

Cellular reproduction is the process that cells follow to make new cells with the same or similar contents as themselves. This process is an essential part of an organism's life. It allows for the organism to grow larger itself, and as it ages, it allows for replacement of dying and damaged cells. The process of cellular

234

reproduction must be accurate and precise. Otherwise, the new cells that are produced will not be able to perform the same functions as the original cell. Mutations can occur in the offspring, which can cause anywhere from minor to severe problems. The two types of cellular reproduction that organisms can use are mitosis or meiosis. **Mitosis** produces daughter cells that are identical to the parent cell and is often referred to as **asexual reproduction**. **Meiosis** has two stages of cell division and produces daughter cells that have a combination of traits from their two parents. It is often referred to as **sexual reproduction**. Humans reproduce by meiosis. During this process, the sperm, the male germ cell, and the egg, the female germ cell, combine and form a parent cell that contains both of their sets of chromosomes. This parent cell then divides into four daughter cells, each with a unique set of traits that came from both of their parents.

In both processes, the most important part of the cell that is copied is the cell's DNA. It contains all of the genetic information for the cell, which leads to its traits and capabilities. Some parts of the cell are copied exactly during cellular reproduction, such as DNA. However, certain other cellular components are synthesized within the new cell after reproduction is complete using the new DNA. For example, the endoplasmic reticulum is broken down during the cell cycle and then newly synthesized after cell division.

Genetics

Humans carry their genetic information on structures called **chromosomes.** Chromosomes are string-like structures made up of nucleic acids and proteins. During the process of reproduction, each parent contributes a gamete that contains twenty-three chromosomes to the intermediate diploid cell. This diploid cell, which contains forty-six chromosomes, replicates itself to produce two diploid cells. These two diploid cells each then split into cells and randomly divide the chromosomes so that each of the four resulting cells contains only twenty-three chromosomes, as each parent gamete does.

Each of the twenty-three chromosomes has between a few hundred and a few thousand genes on it. Each gene contains information about a specific trait that is inherited by the offspring from one of the parents. **Genes** are made up of sequences of DNA that encode proteins and start pathways to express the phenotype that they control. Every gene has two **alleles**, or variations—one inherited from each parent. In most genes, one allele has a more dominant phenotype than the other allele. This means that when both alleles are present on a gene, the dominant phenotype will always be expressed over the recessive phenotype. The recessive phenotype would only be expressed when both alleles present were recessive. Some alleles can have codominance or incomplete dominance. **Codominance** occurs when both alleles are expressed equally when they are both present. For example, the hair color of cows can be red or white, but when one allele for each hair color is present, their hair is a mix of red and white, not pink. **Incomplete dominance** occurs when the presence of two different alleles creates a third phenotype. For example, some flowers can be red or white when the alleles are in duplicate, but when one of each allele is present, the flowers are pink.

Mendel's Laws of Heredity
Gregor Mendel was a monk who came up with one of the first models of inheritance in the 1860s. He is often referred to as the father of genetics. At the time, his theories were largely criticized because biologists did not believe that these ideas could be generally applicable and could also not apply to different species. They were later rediscovered in the early 1900s and given more credence by a group of European scientists. Mendel's original ideas have since been combined with other theories of inheritance to develop the ideas that are studied today.

Between 1856 and 1863, Gregor Mendel experimented with about five thousand pea plants in his garden that had different color flowers to test his theories of inheritance. He crossed purebred white flower and purple flower pea plants and found that the results were not a blend of the two flowers; they were instead all purple flowers. When he then fertilized this second generation of purple flowers with itself, both white flowers and purple flowers were produced, in a ratio of one to three. Although he used different terms at the time, he proposed that the color trait for the flowers was regulated by a gene, which he called a "factor," and that there were two alleles, which he called "forms," for each gene. For each gene, one allele was inherited from each parent. The results of these experiments allowed him to come up with his Principles of Heredity.

There are two main laws that Mendel developed after seeing the results of his experiments. The first law is the **Law of Segregation**, which states that each trait has two versions that can be inherited. In the parent cells, the allele pairs of each gene separate, or segregate, randomly during gamete production. Each gamete then carries only one allele with it for reproduction. During the process of reproduction, the gamete from each parent contributes its single allele to the daughter cell. The second law is the **Law of Independent Assortment**, which states that the alleles for different traits are not linked to one another and are inherited independently. It emphasizes that if a daughter cell selects allele A for gene 1, it does not also automatically select allele A for gene 2. The allele for gene 2 is selected in a separate, random manner.

Mendel theorized one more law, called the **Law of Dominance**, which has to do with the expression of a genotype but not with the inheritance of a trait. When he crossed the purple flower and white flower pea plants, he realized that the purple flowers were expressed at a greater ratio than the white flower pea plants. He hypothesized that certain gene alleles had a stronger outcome on the phenotype that was expressed. If a gene had two of the same allele, the phenotype associated with that allele was expressed. If a gene had two different alleles, the phenotype was determined by the dominant allele, and the other allele, the recessive allele, had no effect on the phenotype.

DNA

DNA is made up of two polynucleotide strands that are linked together and twisted into a double-helix structure. The polynucleotide strands are made up a chain of nucleotides made from four nitrogenous bases, which are adenine, thymine, guanine, and cytosine. Across the two strands, adenine and thymine are paired together, and guanine and cytosine are paired together. This allows for a tight helix to form based on their molecular configurations. DNA contains all of the genetic information of a living organism and provides the protein encoding information on genes. Chromosome replication and cell division start with DNA replication. It is the first step in passing genetic information to subsequent generations. During replication, the double-helix DNA is untwisted and separated. Each strand is replicated and then linked to one of the original strands to form two new DNA molecules. Sometimes, there are errors made in the DNA sequence that codes for a specific gene, causing genetic mutations. If the sequence alteration was originally in the parent gene, it is considered hereditary. If it was developed during the replication process, it is classified as an acquired mutation. Some mutations can cause major phenotypic differences, resulting in developmental problems, while others do not affect development at all. Although they are not very common, mutations are important for the variation of the general population.

Chemistry

Atomic Structure and the Periodic Table

Today's primary model of the atom was proposed by scientist Niels Bohr. Bohr's atomic model consists of a nucleus, or core, which is made up of positively charged protons and neutrally charged neutrons.

Neutrons are theorized to be in the nucleus with the protons to provide "balance" and stability to the protons at the center of the atom. More than 99 percent of the mass of an atom is found in the nucleus. Orbitals surrounding the nucleus contain negatively charged particles called electrons. Since the entire structure of an atom is too small to be seen with the unaided eye, an electron microscope is required for detection. Even with such magnification, the actual particles of the atom are not visible.

An atom has an atomic number that is determined by the number of protons within the nucleus. Some substances are made up of atoms, all with the same atomic number. Such a substance is called an **element.** Using their atomic numbers, elements are organized and grouped by similar properties in a chart called the **Periodic Table.**

The sum of the total number of protons and total number of neutrons in an atom provides the atom's mass number. Most atoms have a nucleus that is electronically neutral, and all atoms of one type have the same atomic number. There are some atoms of the same type that have a different mass number. The variation in the mass number is due to an imbalance of neutrons within the nucleus of the atoms. If atoms have this variance in neutrons, they are called **isotopes.** It is the different number of neutrons that gives such atoms a different mass number.

A concise method of arranging elements by atomic number, similar characteristics, and electron configurations in a tabular format was necessary to represent elements. This was originally organized by scientist Dmitri Mendeleev using the Periodic Table. The vertical columns on the Periodic Table are called **groups** and are sorted by similar chemical properties/characteristics, such as appearance and reactivity. This is observed in the shiny texture of metals, the softness of post-transition metals, and the high melting points of alkali earth metals. The horizontal rows on the Periodic Table are called **periods** and are arranged by electron valance configurations.

The Periodic Table of the Elements

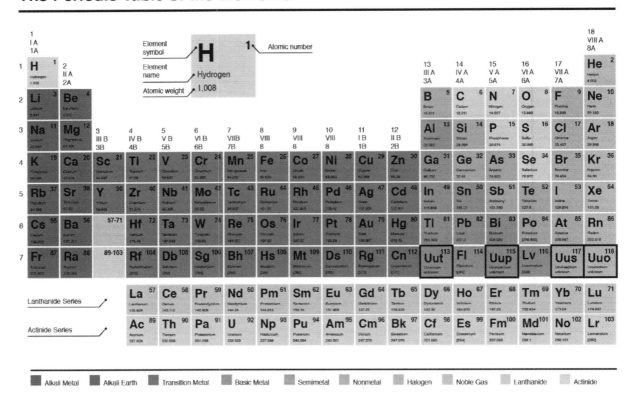

Elements are set by ascending atomic number, from left to right. The number of protons contained within the nucleus of the atom is represented by the atomic number. For example, hydrogen has one proton in its nucleus, so it has an atomic number of 1.

Since isotopes can have different masses within the same type of element, the **atomic mass** of an element is the average mass of all the naturally occurring atoms of that given element. Atomic mass is calculated by finding the relative abundance of isotopes that might be used in chemistry, or by adding the number of protons and neutrons of an atom together. For example, the atomic number of chlorine is 35 on the Periodic Table. However, the atomic mass of chlorine is 35.5 atomic mass units (amu). This discrepancy occurs because a large number of chlorine isotopes (meaning, instead of 35 neutrons, a chlorine nucleus might have 36 neutrons) exist in nature. The average of all the atomic masses turns out to be 35.5 amu, which is slightly higher than chlorine's listed number on the Periodic Table. Carbon has an atomic number of 12, but its atomic mass is 12.01 amu because there are not as many naturally occurring isotopes to raise the average number, as observed with chlorine.

Chemical Reactions

A chemical reaction is a process that involves a change in the molecular arrangement of a substance. Generally, one set of chemical substances, called the **reactants**, is rearranged into a different set of chemical substances, called the **products**, by the breaking and re-forming of bonds between atoms. In a chemical reaction, it is important to realize that no new atoms or molecules are introduced. The products are formed solely from the atoms and molecules that are present in the reactants. These can involve a change in state of matter as well. Making glass, burning fuel, and brewing beer are all examples of chemical reactions.

Generally, chemical reactions are thought to involve changes in positions of electrons with the breaking and re-forming of chemical bonds, without changes to the nucleus of the atoms. Three of the main types of chemical reactions are combination, decomposition, and combustion.

Combination
In combination reactions, two or more reactants are combined to form one more complex, larger product. The bonds of the reactants are broken, the elements are arranged, and then new bonds are formed between all of the elements to form the product. It can be written as A + B → C, where A and B are the reactants and C is the product. An example of a combination reaction is the creation of iron(II) sulfide from iron and sulfur, which is written as:

$$8Fe + S_8 \rightarrow 8FeS$$

Decomposition
Decomposition reactions are almost the opposite of combination reactions. They occur when one substance is broken down into two or more products. The bonds of the first substance are broken, the elements are rearranged, and then the elements are bonded together in new configurations to make two or more molecules. These reactions can be written as C → B + A, where C is the reactant and A and B are the products. An example of a decomposition reaction is the electrolysis of water to make oxygen and hydrogen gas, which is written as:

$$2H_2O \rightarrow 2H_2 + O_2$$

Combustion
Combustion reactions are a specific type of chemical reaction that involves oxygen gas as a reactant. This mostly involves the burning of a substance. The combustion of hexane in air is one example of a combustion reaction. The hexane gas combines with oxygen in the air to form carbon dioxide and water. The reaction can be written as:

$$2C_6H_{14} + 17O_2 \rightarrow 12CO_2 + 14H_2O$$

Stoichiometry

Stoichiometry uses proportions based on the principles of the conservation of mass and the conservation of energy. It deals with first balancing the chemical (and sometimes physical) changes in a reaction and then finding the ratios of the reactants used and what products resulted. Just as there are different types of reactions, there are different types of stoichiometry problems. Different reactions can involve mass, volume, or moles in varying combinations. The steps to solve a stoichiometry problem are: first, balance the equation; next, find the number of total products; and finally, calculate the desired information regarding molar mass, percent yield, etc.

The **molar mass** of any substance is the measure of the mass of one mole of that substance. For pure elements, the molar mass is also referred to as the **atomic mass unit (amu)** for that substance. In compounds, molar mass is calculated by adding the molar mas of each substance in the compound. For example, the molar mass of carbon can be found on the Periodic Table as 12.01 g/mol, while finding the molar mass of water (H_2O) requires a bit of calculation:

$$the\ molar\ mass\ of\ hydrogen = 1.10\ g/mol \times 2 = 2.02\ g/mol$$

$$+\ the\ molar\ mass\ of\ oxygen = 16.0\ g/mol = 16.00\ g/mol$$

$$2.02 + 16.00 = 18.02\ g/mol$$

To determine the percent composition of a compound, the individual molar masses of each component need to be divided by the total molar mass of the compound and then multiplied by 100. For example, to find the percent composition of carbon dioxide (CO_2), the molar mass of CO_2 must be calculated first:

$$the\ molar\ mass\ of\ carbon = 12.01\ g/mol = 12.01\ g/mol$$

$$+\ the\ molar\ mass\ of\ oxygen = 16.0\ g/mol \times 2 = 32.00\ g/mol$$

$$12.01 + 32.00 = 18.02\ g/mol$$

Next, take each individual mass, divide by the total mass of the compound, and then multiply by 100 to get the percent composition of each component of the compound:

$$\frac{12.01\ g/mol}{44.01\ g/mol} = 0.2729 \times 100 = 27.29\%\ carbon$$

$$\frac{32.00\ g/mol}{44.01\ g/mol} = 0.7271 \times 100 = 72.71\%\ oxygen$$

A quick check in the addition of the percentages should always total 100%.

If an example provides the basis for an equation, the equation would first need to be balanced to calculate any proportions. For example, if 15 g of C_2H_6 reacts with 64 g of O_2, how many grams of CO_2 will be formed?

First, write the chemical equation:

$$C_2H_6\ +\ O_2 \rightarrow CO_2\ +\ H_2O$$

Next, balance the equation:

$$2\ C_2H_6\ +\ 7\ O_2 \rightarrow 4\ CO_2\ +\ 6\ H_2O$$

Then, calculate the desired amount based on the beginning information of 15 g of C_2H_6:

$$15\ g\ C_2H_6 \times \frac{1\ mole\ C_2H_6}{30\ g\ C_2H_6} \times \frac{4\ moles\ CO_2}{2\ moles\ C_2H_6} \times \frac{44\ g\ CO_2}{1\ mole\ CO_2} = 44\ g\ CO_2$$

To check that this would be the smaller amount (or how much until one of the reactants is used up, thus ending the reaction), the calculation would need to be done for the 64 g of O_2:

$$64 \; g \; O_2 \; \times \; \frac{1 \; mole \; O_2}{32 \; g \; O_2} \; \times \; \frac{4 \; moles \; CO_2}{7 \; moles \; O_2} \; \times \; \frac{44 \; g \; CO_2}{1 \; mole \; CO_2} = 50.5 \; g \; CO_2$$

The yield from the C_2H_6 is smaller, so it would be used up first, ending the reaction. This calculation would determine the maximum amount of CO_2 that could possibly be produced.

Acids and Bases

If something has a sour taste, it is considered acidic, and if something has a bitter taste, it is considered basic. Acids and bases are generally identified by the reaction they have when combined with water. An acid will increase the concentration of hydrogen ions (H^+) in water, and a base will increase the concentration of hydroxide ions (OH^-). Other methods of identification with various indicators have been designed over the years.

To better categorize the varying strength levels of acids and bases, the pH scale is employed. The **pH scale** is a logarithmic (base 10) grading applied to acids and bases according to their strength. The pH scale contains values from 0 through 14 and uses 7 as neutral. If a solution registers below a 7 on the pH scale, it is considered an acid. If a solution registers higher than a 7, it is considered a base. To perform a quick test on a solution, litmus paper can be used. A base will turn red litmus paper blue, and an acid will turn blue litmus paper red. To gauge the strength of an acid or base, a test using phenolphthalein can be administered. An acid will turn red phenolphthalein to colorless, and a base will turn colorless phenolphthalein to pink. As demonstrated with these types of tests, acids and bases neutralize each other. When acids and bases react with one another, they produce salts (also called **ionic substances**).

Acids and bases have varying strengths. For example, if an acid completely dissolves in water and ionizes, forming an H^+ and an anion, it is considered a strong acid. There are only a few common strong acids, including sulfuric (H_2SO_4), hydrochloric (HCl), nitric (HNO_3), hydrobromic (HBr), hydroiodic (HI), and perchloric ($HClO_4$). Other types of acids are considered weak. An easy way to tell if something is an acid is by looking for the leading "H" in the chemical formula.

A base is considered strong if it completely dissociates into the cation of OH^-, including sodium hydroxide (NaOH), potassium hydroxide (KOH), lithium hydroxide (LiOH), cesium hydroxide (CsOH), rubidium hydroxide (RbOH), barium hydroxide ($Ba(OH)_2$), calcium hydroxide ($Ca(OH)_2$), and strontium hydroxide ($Sr(OH)_2$). Just as with acids, other types of bases are considered weak. An easy way to tell if something is a base is by looking for the "OH" ending on the chemical formula.

In pure water, autoionization occurs when a water molecule (H_2O) loses the nucleus of one of the two hydrogen atoms to become a hydroxide ion (OH^-). The nucleus then pairs with another water molecule to form hydronium (H_3O^+). This autoionization process shows that water is **amphoteric,** which means it can react as an acid or as a base.

Pure water is considered neutral, but the presence of any impurities can throw off this neutral balance, causing the water to be slightly acidic or basic. This can include the exposure of water to air, which can introduce carbon dioxide molecules to form carbonic acid (H_2CO_3), thus making the water slightly acidic. Any variation from the middle of the pH scale (7) indicates a non-neutral substance.

Earth/Space Sciences

The Structure of the Earth System

The structure of the Earth has many layers. Starting with the center, or the **core,** the Earth comprises two separate sections: the inner core and the outer core. The innermost portion of the core is a solid center consisting of approximately 760 miles of iron. The outer core is slightly less than 1400 miles in thickness and consists of a liquid nickel-iron alloy. The next section out from the core also has two layers. This section is the **mantle,** and it is split into the lower mantle and the upper mantle. Both layers of the mantle consist of magnesium and iron, and they are extremely high in temperature. This hot temperature causes the metal contained in the lower mantle to rise and then cool slightly as it reaches the upper mantle. Once the metal begins to cool, it falls back down toward the lower mantle, restarting the whole process again. The motion of rising and falling with in the layers of the mantle is the cause of plate tectonics and movement of the outermost layer of the Earth. The outermost layer of the Earth is called the **crust.** Movements between the mantle and the crust create effects such as earthquakes and volcanoes.

The Earth's atmosphere is composed of five primary layers. These layers consist of gases that are held in place by the force due to the effects of Earth's gravity. The breakdown of the gases includes mostly nitrogen (78 percent), and less than one-quarter oxygen (21 percent); the rest is other gases (approximately 1 percent in total). Nearly all of the weather and clouds exist in the closest layer to the Earth's surface, called the **troposphere**. After this first layer, the atmosphere thins out because each layer is farther away from the surface of the Earth and therefore less impacted by the effects of gravity. This thinning of layers continues until the atmosphere meets space. The layer closest to the Earth contains half of the actual atmosphere, reaches approximately twelve miles vertically, and is heated by the surface of the Earth. As this first layer becomes farther away from the surface, it cools off and contains water vapor, which forms clouds.

The second layer of the atmosphere from the surface of the Earth, called the **stratosphere,** is very dry and contains ozone, which helps to absorb potentially harmful wavelengths from solar radiation. The second layer is also where airplanes and weather balloons fly.

The third layer from the surface of the Earth, the **mesosphere,** is the coldest. This layer is where meteors will burn up if they enter the atmosphere. The fourth layer, the **thermosphere,** has very low air density and is where the space shuttle and the International space station orbit. The fifth, and final thin layer, called the **exosphere,** consists of hydrogen and helium.

Processes of the Earth System

The cycle of water on the Earth is called the **hydrologic cycle.** This cycle involves the water from the surface of the Earth evaporating from the oceans, lakes, and rivers, into the air. This evaporating water cools as it rises in the air. As the cooling occurs, the water condenses into clouds contained within the atmosphere, which eventually allow the water to return to the surface of the Earth in some form of precipitation. Precipitation can occur as snow, rain, hail, or sleet. Areas experiencing drought receive little or no precipitation; therefore, little to no water is available to run off into the surrounding bodies of water. The hydrologic cycle becomes imbalanced and often dormant in affected regions.

The terms *weather* and *climate* are often mistakenly interchanged, even though they describe different phenomena on Earth. **Weather** on Earth is constantly changing, while **climate** describes a long-term state. Factors that can affect the weather include latitude, elevation, wind, proximity to a large body of water, and ocean currents. Latitude influences weather based upon distance from the equator; for instance, the sections of the Earth nearest to the equator receive more direct sunlight from the positioning of the Earth on its axis, and therefore are warmer. The higher the elevation of a location above sea level, the colder the temperature. Wind, resulting from the Earth's rotation (trade winds), can affect the temperatures of the surrounding areas. Large bodies of water can store heat, which influences the weather of surrounding areas and the effect ocean currents have on the rising and falling of warm air. This phenomenon occurs around lakes and surrounding areas; at times, for instance, areas on one side of a lake will experience heavier amounts of snowfall due to what is called "lake effect snow." This occurs when the lake is at a higher temperature than the atmosphere, and winds, picking up some of this energy, which causes a larger amount of snowfall on the other side of the lake.

Earth History

Scientists have theorized that the Earth was formed by the "Big Bang" approximately ten to fifteen billion years ago. The **Big Bang** describes an initial expansion of the universe from a high-density, high-temperature, state. The entire universe has been expanding ever since the occurrence of the Big Bang; this expansion of gases and matter into condensed clouds formed the galaxies, and more specifically, the components of the solar system. The progression in the stages of the formation of the universe is referred to as **evolution**. The term *evolution* is also used to describe the development of organisms in a biological progress. One method used for measuring the age of the universe is the measurement of the decay of present radioactive materials contained in rocks. This can also be used for dating the age of rocks and minerals on the Earth. The process of dating such substances involves calculating the **half-life** (all radioactive materials decay at an exponential rate referred to as its half-life) of these radioactive materials and determining when they would have come into existence.

The study of the existing rocks on Earth are an easy way to record the types of elements and minerals that still exist from the beginnings of the universe. Tracking how much of the elements and minerals still exist, despite their decay, is how dating is computed. A comparison to known dates and phases already established on Earth is used to apply new timeframes to these materials. Often, there are fossils preserved in the rock on the Earth, which can make dating such materials easier, since most fossils are from organisms that have existed within a confined window of time.

The study of the age of the Earth and the materials on the Earth is called **paleontology,** a science often considered a bridge between biology and geology. Not only does paleontology study the existence and history of fossils and materials on Earth, it attempts to determine the causes of their existence. The best way to determine the date of an organic object is through **carbon dating**. This a process that takes a sample of the Carbon 14 isotope in question and compares it to the known values of Carbon 12, in order to determine the amount of carbon that has decayed into nitrogen. The discrepancy between the two values provides a rather accurate timeframe from which to trace the organic object's existence. Another element used for dating nonorganic objects is uranium. Uranium has a consistent and predictable decay rate to lead that can be used for comparison.

Dating Rocks

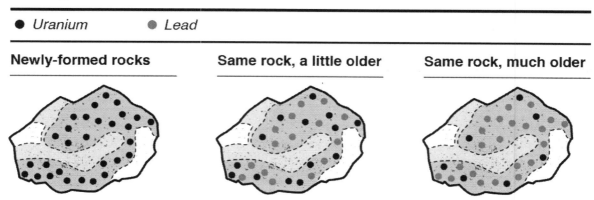

● *Uranium*   ● *Lead*

Newly-formed rocks   **Same rock, a little older**   **Same rock, much older**

Earth and the Universe

The **universe** is defined as the largest entity made by space and time, and it includes all of the smaller entities contained within it. The largest known systems within the universe are known as **galaxies**. Scientists believe there are over 100 billion galaxies within the universe. Galaxies can be made up of different gases, cosmic debris, stars, planets, moons, black holes, and other bits of matter. Billions of these entities can make up a single galaxy. They primarily appear in either spiral or elliptical shapes. Galaxies that cannot be categorized as these shapes are known as irregularly-shaped galaxies. These different shapes are made based on the masses of the various entities within the galaxy. The gravitational pull of these objects upon one another creates not only the shape of the galaxy, but also influences how other systems operate within the galaxy. Galaxies can be hundreds of thousands to millions of light years across in size.

Within galaxies, the next largest system is a **solar system**. These systems contain of one or more stars with a massive gravitational pull. Various planets, moons, and debris orbit the star or stars as a result. Stars form because of the gravitational collapse of explosive elements (primarily hydrogen and helium). The center of the star continues in a state of thermonuclear fusion that creates the star's internal gravitational pull. Over many thousands of years, the central thermonuclear activity will slowly cease until the star collapses, resulting in the absorption of orbiting bodies. The Earth is part of a solar system in the Milky Way galaxy. Its solar system consists of one sun, eight planets, their moons, and other cosmic bodies, such as meteors and asteroids (including an asteroid belt that orbits the Sun in a full ring). In order of proximity to the Sun, the planets in this solar system are Mercury, Venus, Earth, Mars, Jupiter, Saturn, Uranus, and Neptune. Beyond Neptune is a dwarf planet named Pluto, and at least four other known dwarf planets. While the Earth has only a single moon, most other planets have multiple moons (although Mercury and Venus have none). Based on their mass, moons, and position from the Sun, each planet takes a different period to orbit the Sun and rotate upon its axis. The Earth takes 24 hours to rotate on its axis and 365 days to orbit the Sun.

Solar eclipses occur on Earth when the moon's orbit appears to cross between the Earth and the Sun, while lunar eclipses occur when the Earth's orbit crosses between the Sun and the moon. Additionally, the moon's gravitational pull on the Earth impacts the water on the planet, causing high and low tides throughout the day. The difference in tidal height varies based on the location of the moon in its orbit around the Earth (as well as different features of the water body and shoreline).

The positioning of the moon between the Earth and the Sun at different positions in the Earth's orbit results in the phases of the moon. While a common misconception is that the phases are simply caused from the shadow the Earth casts on the moon, the different phases we see are actually due to the position of the moon relative to the Sun. A portion of the moon that is not visible or shadowed is turned away from the Sun. At all times, half of the moon is illuminated while half is shadowed, yet our perception of different phases is caused by the moon's position relative to us on Earth.

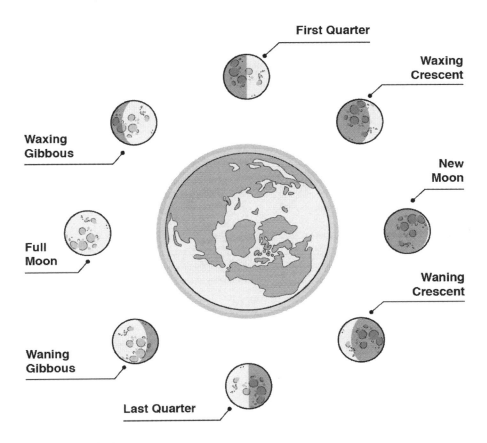

Earth Patterns, Cycles, and Change

There are three main sources that change the shape of the Earth's surface: they include weathering, erosion, and plate tectonics.

Weathering is the natural physical and chemical breakdown of rocks and the minerals within them. Some of the more common agents of weathering include changes in temperature, acids and acid rain, salts, wind, water, ice, plants, and animals. For example, water can seep into porous layers of bedrock. Because ice is less dense than water, when water freezes, it expands. Thus, the freeze-thaw cycle can cause rocks to crack and split into pieces. Once weathering has occurred, by any of its possible means, erosion can act upon the dissolved rocks and minerals and transport them away from their original site.

Erosion is caused by wind, water, and ice, and is considered a natural process that can be expedited by the effects of humans. Deforestation and overgrazing can expose soil, leading to extra erosion. In many areas, soil that is exposed can be carried away by strong winds and this changes the shape of the Earth's surface.

Glaciers contribute to erosion. They are so massive and heavy that their weight dislodges rocks. This, along with water melting and freezing in the cracks of bedrock, moves rocks that a glacier crushes back into fine rock "flour" as it slowly progresses along its path.

The Rock Cycle

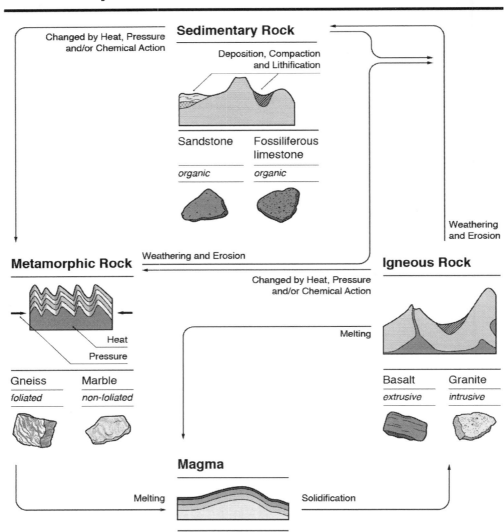

Often, **magma** (hot fluid below the earth's crust) from the Earth's mantle can find its way through cracks or boundaries between the plates in weak spots in the Earth's surface. This causes volcanoes, which can be a slow seeping action, called **shield domes**, or an explosive action at the surface, called **composite cones**. Both types of volcanoes cause corrosion of certain portions of land, but they can also add to the structure of the surface due to depositing amounts of silica upon the magma cooling, when exposed to a lower temperature atmosphere. Other types of volcanoes include the **cinder cones**, where the lava can burst up and then quickly cool and collect on the sides and lava domes, where the lava will pile up near the vent to form a steep side. There could also be a crack over the top of a lava flow, which is called a **fissure volcano**, and a volcano that keeps recreating itself within its original vent, which is called a **caldera volcano**.

Types of Volcano

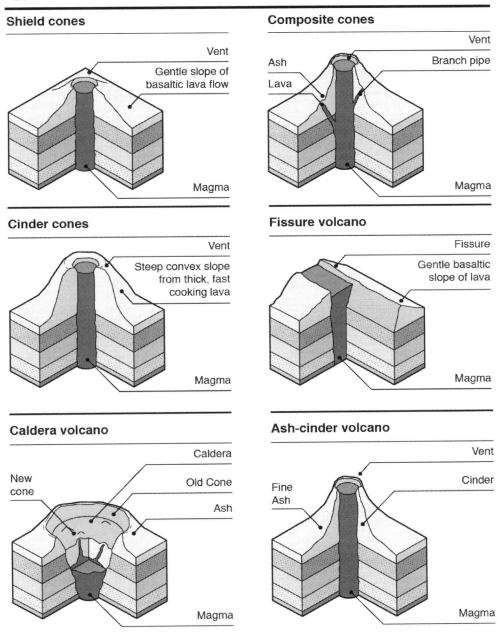

Shield cones
- Vent
- Gentle slope of basaltic lava flow
- Magma

Composite cones
- Vent
- Ash
- Lava
- Branch pipe
- Magma

Cinder cones
- Vent
- Steep convex slope from thick, fast cooking lava
- Magma

Fissure volcano
- Fissure
- Gentle basaltic slope of lava
- Magma

Caldera volcano
- New cone
- Caldera
- Old Cone
- Ash
- Magma

Ash-cinder volcano
- Vent
- Fine Ash
- Cinder
- Magma

Similar to how molecules transfer heat, convection currents within the Earth's mantle cause cooling and heating of the magma, making it move in circular motions. These convection currents cause constant movement in the plates on the Earth's crust. It is **plate tectonics**—the movement of the plates past one another—that causes earthquakes. Earthquakes can be destructive by creating **craters**, which swallow up portions of the Earth and by forming **ridges**, where parts of the Earth pile up on each other. Both actions are transforming to the Earth's surface.

Science as a Human Endeavor, a Process, and a Career

People of all cultures across the world utilize science to explore questions and find solutions to problems. The systematic process of designing, conducting, and analyzing experiments is universally known and respected. These processes are time consuming and require specific knowledge and skills. Therefore, the pursuit of science is its own career path, with many smaller paths for each respective area of study (i.e., life sciences, chemical sciences, physical sciences). Of course, each of those paths splits into even more refined areas and requires much study and dedication. Men and women alike pursue the questions of the world around us—some are driven by pure curiosity, and others are compelled by finding a faster or more economical way of performing a task or producing an object.

Not all ideas, methods, or results are popular or accepted by those in society. Thus, the pursuit of science is often riddled with controversy. This has been an underlying theme since the early days of astronomical discovery. Copernicus was excommunicated from his religious establishment when he announced his belief that the Sun, not the Earth, was the controlling body of the heavens known to humans at the time. Despite Copernicus having documented observations and calculations, those opposed to his theory could not be convinced; therefore, Copernicus experienced great ridicule and suffering due to his scientific research and assertions. In addition, other scientists have faced adverse scrutiny for their assertions, including Galileo, Albert Einstein, and Stephen Hawking. In each case, logical thoughts, observations, and calculations were used to demonstrate their ideas, yet opposition to their scientific beliefs still exists.

The possibilities for careers involving science range from conducting research, to the application of science and research (engineering), to academia (teaching). All of these avenues require intensive study and a thorough understanding of respective branches of science and their components. An important factor of studying and applying science is being able to concisely and accurately communicate knowledge to other people. Many times, this is done utilizing mathematics or demonstration. The necessity of communicating ideas, research, and results brings people from all nationalities together. This often lends to different cultures finding common ground for research and investigation, and it opens lines of communication and cooperation.

Using Resource and Research Material in Science

Part of the process of scientific inquiry is researching a problem or question. Before an experiment can be designed, proper research must be conducted into the question at hand. The initial question needs to be well formed and based on logical reasoning. A literature review should be conducted on existing material pertaining to the subject in question, and confirmation of any experimentation on the question that has been conducted previously. If prior experimentation exists, what were the results obtained and were any conclusions drawn from those results? In addition, research must be done on all possible information regarding the initial question, the experiment, how to investigate the question, and what tools will be necessary to draw conclusions and explain any findings. Just as an experiment must be unbiased, so should be any research regarding the experiment. All sources of information need to be proven reliable and accredited. For instance, a person's account of their opinion on a situation does not constitute as a valid source for research; sources s be free of opinion or speculation.

During experimentation, research must be conducted with appropriate mechanisms for observation and measurement. Knowing the proper tools and units for accurately measuring a volume or a mass is a fundamental skill of research. Researchers also need to be held to a standard of ethics and honesty. The independent repetition of an experiment helps to ensure this level of accountability. Often, the most reliable resources are those of accredited experimenters, universities, and other research laboratories. In order for such sources to publish information, they need to demonstrate strict adherence to the scientific methods, precise measurements for observations, and specific mathematical reporting.

It is often common for different scientists in the same place, or even varying countries, to be conducting experiments to test the same hypothesis. This does not always lead to a race to see who finishes first, but it can lead to cooperative research and shared acclaim if the results prove successful. Awards for research, discoveries, and scientific application are often used by the scientific community to show appreciation for advancements in science.

The Unifying Processes of Science

Following the scientific method and keeping to the standards of proper research and reporting, lends to easier communication of data and results. When information is able to be conveyed to multiple audiences in a manner of common understanding (i.e., mathematics), it increases the possibilities for the use of such information. Enabling other scientists understand an idea can also lead to further experimentation and discovery in that area, which, in turn, leads to the further organization of information and a deeper understanding of our universe. It is more systematic to group, sort, and organize information for commonalities in order to increase this understanding. This organization of groups can also serve as a reference point when attempting to identify other members of that group, on introduce new members to it. For instance, a newly discovered type of rock can be compared to known types of rocks and then better categorized as to its uses or properties based upon how it appears and responds to experiment when measured against known rocks.

This occurs regularly when varying crystal rock structures are developed for use in super-cooled or super-conductive experiments. Certain properties are more useful with regard to conduction and strength. In order to have knowledge and access to this type of variation of information, societies are formed and people from all over the world find ways to communicate and share in the scientific endeavor. The communication of research can further questions and explorations across the world, and this common goal of reaching new discoveries or uses for the application of science can bring people together. Often, the quest for scientific discovery is spawned by competition or the race to create something before another society or country. Examples of this include the race to explore space, the race for nuclear armaments, and the race to create and cure strains of deadly bacteria. In these situations, the urge to push scientific discovery ahead may not be for the most humanitarian motives; however, often such research prompts results in accidental discoveries that can solve other problems. The discoveries of vaccinations, stronger materials such as plastics, and cleaner forms of energy through superconducting crystals, have all been accidental discoveries along the way of competitive scientific research. Whatever the motive for scientific discovery, it can be a seen as a common drive across many nations with a potential to create unity through its demand for structure and organization.

Physics

Nature of Motion

Cultures have been studying the movement of objects since ancient times. These studies have been prompted by curiosity and sometimes by necessity. On Earth, items move according to guidelines and have motion that is fairly predictable. To understand why an object moves along its path, it is important to understand what role forces have on influencing its movements. The term **force** describes an outside influence on an object. Force does not have to refer to something imparted by another object. Forces can act upon objects by touching them with a push or a pull, by friction, or without touch like a magnetic force or even gravity. Forces can affect the motion of an object.

To study an object's motion, it must be located and described. When locating an object's position, it can help to pinpoint its location relative to another known object. Comparing an object with respect to a known object is referred to as **establishing a frame of reference**. If the placement of one object is known, it is easier to locate another object with respect to the position of the original object.

Motion can be described by following specific guidelines called **kinematics.** Kinematics use mechanics to describe motion without regard to the forces that are causing such motions. Specific equations can be used when describing motions; these equations use time as a frame of reference. The equations are based on the change of an object's position (represented by x), over a change in time (represented by Δt). This describes an object's velocity, which is measured in meters/second (m/s) and described by the following equation:

$$v = \frac{\Delta x}{\Delta t} = \frac{x_f - x_i}{\Delta t}$$

Velocity is a vector quantity, meaning it measures the magnitude (how much) and a direction (that the object is moving). Both of these components are essential to understanding and predicting the motion of objects. The scientist Isaac Newton did extensive studies on the motion of objects on Earth and came up with three primary laws to describe motion:

Law 1: An object in motion tends to stay in motion unless acted upon by an outside force. An object at rest tends to stay at rest unless acted upon by an outside force (also known as the **law of inertia**).

For example, if a book is placed on a table, it will stay there until it is moved by an outside force.

Law 2: The force acting upon an object is equal to the object's mass multiplied by its acceleration (also known as **F = ma**).

For example, the amount of force acting on a bug being swatted by a flyswatter can be calculated if the mass of the flyswatter and its acceleration are known. If the mass of the flyswatter is 0.3 kg and the acceleration of its swing is 2.0 m/s², the force of its swing can be calculated as follows:

$$m = 0.3\,kg$$
$$a = 2.0\,m/s^2$$
$$F = m \times a$$
$$F = (0.3) \times (2.0)$$
$$F = 0.6\,N$$

Law 3: For every action, there is an equal and opposite reaction.

For example, when a person claps their hands together, the right hand feels the same force as the left hand, as the force is equal and opposite.

Another example is if a car and a truck run head-on into each other, the force experienced by the truck is equal and opposite to the force experienced by the car, regardless of their respective masses or velocities. The ability to withstand this amount of force is what varies between the vehicles and creates a difference in the amount of damage sustained.

Newton used these laws to describe motion and derive additional equations for motion that could predict the position, velocity, acceleration, or time for objects in motion in one and two dimensions. Since all of Newton's work was done on Earth, he primarily used Earth's gravity and the behavior of falling objects to design experiments and studies in free fall (an object subject to Earth's gravity while in flight). On Earth, the acceleration due to the force of gravity is measured at 9.8 meters per second² (m/s²). This value is the same for anything on the Earth or within Earth's atmosphere.

Kinetic Energy and Potential Energy

There are two main types of energy. The first type is called **potential energy** (or **gravitational potential energy**), and it is stored energy, or energy due to an object's height from the ground.

The second type is called **kinetic energy**. Kinetic energy is the energy of motion. If an object is moving, it will have some amount of kinetic energy.

For example, if a roller-coaster car is sitting on the track at the top of a hill, it would have all potential energy and no kinetic energy. As the roller coaster travels down the hill, the energy converts from potential energy into kinetic energy. At the bottom of the hill, where the car is traveling the fastest, it would have all kinetic energy and no potential energy.

Another measure of energy is the **total mechanical energy** in a system. This is the sum (or total) of the potential energy plus the kinetic energy of the system. The total mechanical energy in a system is always conserved. The amounts of the potential energy and kinetic energy in a system can vary, but the total mechanical energy in a situation would remain the same.

The equation for the mechanical energy in a system is as follows:

$$ME = PE + KE$$

$$(Mechanical\ Energy\ =\ Potential\ Energy\ +\ Kinetic\ Energy)$$

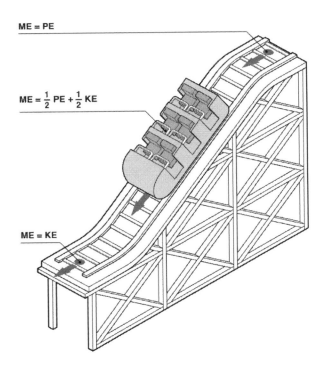

251

Energy can transfer or change forms, but it cannot be created or destroyed. This transfer can take place through waves (including light waves and sound waves), heat, impact, etc.

There is a fundamental law of **thermodynamics** (the study of heat and movement) called **conservation of energy.** This law states that energy cannot be created or destroyed, but rather energy is transferred to different forms involved in a process. For instance, a car pushed beginning at one end of a street will not continue down that street forever; it will gradually come to a stop some distance away from where it was originally pushed. This does not mean the energy has disappeared or has been exhausted; it means the energy has been transferred to different mediums surrounding the car. Some of the energy is dissipated by the frictional force from the road on the tires, the air resistance from the movement of the car, the sound from the tires on the road, and the force of gravity pulling on the car. Each value can be calculated in a number of ways, including measuring the sound waves from the tires, the temperature change in the tires, the distance moved by the car from start to finish, etc. It is important to understand that many processes factor into such a small situation, but all situations follow the conservation of energy.

Just like the earlier example, the roller coaster at the top of a hill has a measurable amount of potential energy; when it rolls down the hill, it converts most of that energy into kinetic energy. There are still additional factors such as friction and air resistance working on the coaster and dissipating some of the energy, but energy transfers in every situation.

There are six basic machines that utilize the transfer of energy to the advantage of the user. These machines function based on an amount of energy input from the user and accomplish a task by distributing the energy for a common purpose. These machines are called **simple machines** and include the lever, pulley, wedge, inclined plane, screw, and wheel and axle.

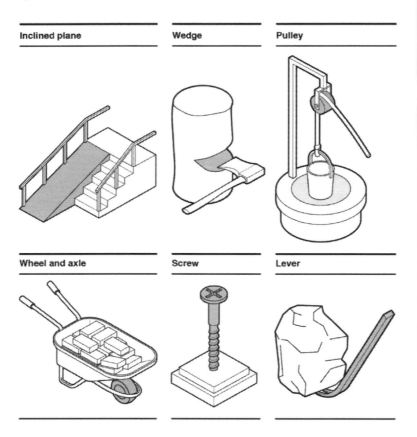

Inclined plane

Wedge

Pulley

Wheel and axle

Screw

Lever

The use of simple machines can help by requiring less force to perform a task with the same result. This is referred to as a **mechanical advantage**.

For example, if a father is trying to lift his child into the air with his arms to pick an apple from a tree, it would require less force to place the child on one end of a teeter totter and push the other end of the teeter totter down to elevate the child to the same height to pick the apple. In this example, the teeter totter is a lever.

Universal Gravitation

Every object in the universe that has mass causes an attractive force to every other object in the universe. The amount of attractive force depends on the masses of the two objects in question and the distance that separates the objects. This is called the **law of universal gravitation** and is represented by the following equation:

$$F = G\frac{m_1 m_2}{r^2}$$

In this equation, the force, F, between two objects, m_1 and m_2, is indirectly proportional to the square of the distance separating the two objects. A general gravitational constant:

$$G\ (6.67 \times 10^{-11}\ \frac{N \cdot m^2}{kg^2})$$

is multiplied by the equation. This constant is quite small, so for the force between two objects to be noticeable, they must have sizable masses.

To better understand this on a large scale, a prime representation could be viewed by satellites (planets) in the solar system and the effect they have on each other. All bodies in the universe have an attractive force between them. This is closely seen by the relationship between the Earth and the moon. The Earth and the moon both have a gravitational attraction that affects each other. The moon is smaller in mass than the Earth; therefore, it will not have as big of an influence as the Earth has on it. The attractive force from the moon is observed by the systematic push and pull on the water on the face of the Earth by the rotations the moon makes around the Earth. The tides in oceans and lakes are caused by the moon's gravitational effect on the Earth. Since the moon and the Earth have an attractive force between them, the moon pulls on the side of the Earth closest to the moon, causing the waters to swell (high tide) on that side and leave the ends 90 degrees away from the moon, causing a low tide there. The water on the side of the Earth farthest from the moon experiences the least amount of gravitational attraction so it collects on that side in a high tide.

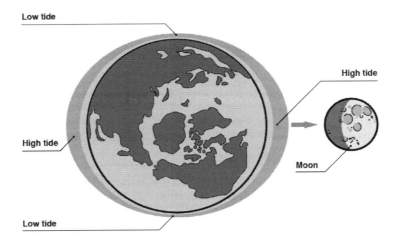

The universal law of gravitation is taken primarily from the works of Johannes Kepler and his laws of planetary motion. These include the fact that the paths of the orbits of the planets are not perfect circles, but ellipses, around the sun. The area swept out between the planet and the sun is equal at every point in the orbit due to fluctuation in speed at different distances. Finally, the period (T) of a planet's motion squared is inversely proportional to the distance (r) between that planet and the sun cubed.

$$\frac{T^2}{r^3}$$

Sir Isaac Newton used this third law and applied it to the idea of forces and their effects on objects. The effect of the gravitational forces of the moon on the Earth are noted in the tides, and the effect of the forces of the Earth on the moon are noted in the fact that the moon is caught in an orbit around the earth. Since the moon is traveling at a velocity tangent to its orbit around the Earth and the Earth keeps attracting it in, the moon does not escape and does not crash into the Earth. The moon will continue this course due to the attractive gravitational force between the Earth and the moon. Albert Einstein later applied Newton's adaptation of Kepler's laws. Einstein was able to develop a more advanced theory, which could explain the motions of all the planets and even be applied beyond the solar system. These theories have also been beneficial for predicting behaviors of other objects in the Earth's atmosphere, such as shuttles and astronauts.

Waves and Sound

Mechanical waves are a type of wave that pass through a medium (solid, liquid, or gas). There are two basic types of mechanical waves: longitudinal and transverse.

A **longitudinal wave** has motion that is parallel to the direction of the wave's travel. It can best be shown by compressing one side of a tethered spring and then releasing that end. The movement travels in a bunching and then unbunching motion, across the length of the spring and back until the energy is dissipated through noise and heat.

A **transverse wave** has motion that is perpendicular to the direction of the wave's travel. The particles on a transverse wave do not move across the length of the wave but oscillate up and down to create the peaks and troughs observed on this type of wave.

A wave with a mix of both longitudinal and transverse motion can be seen through the motion of a wave on the ocean, with peaks and troughs, oscillating particles up and down.

Mechanical waves can carry energy, sound, and light. Mechanical waves need a medium through which transport can take place. However, an electromagnetic wave can transmit energy without a medium, or in a vacuum.

Sound travels in waves and is the movement of vibrations through a medium. It can travel through air (gas), land, water, etc. For example, the noise a human hears in the air is the vibration of the waves as they reach the ear. The human brain translates the different frequencies (pitches) and intensities of the vibrations to determine what created the noise.

A tuning fork has a predetermined frequency because of the size (length and thickness) of its tines. When struck, it allows vibrations between the two tines to move the air at a specific rate. This creates a specific tone (or note) for that size of tuning fork. The number of vibrations over time is also steady for that tuning fork and can be matched with a **frequency** (the number of occurrences over time). All sounds heard by the human ear are categorized by using frequency and measured in **hertz** (the number of cycles per second).

The intensity (or loudness) of sound is measured on the Bel scale. This scale is a ratio of one sound's intensity with respect to a standard value. It is a logarithmic scale, meaning it is measured by factors of ten. But the value that is 1/10 of this value, the decibel, is the measurement used more commonly for the intensity of pitches heard by the human ear.

The **Doppler effect** applies to situations with both light and sound waves. The premise of the Doppler effect is that, based on the relative position or movement of a source and an observer, waves can seem shorter or longer than they are. When the Doppler effect is experienced with sound, it warps the noise being heard by the observer by making the pitch or frequency seem shorter or higher as the source is approaching and then longer or lower as the source is getting farther away. The frequency and pitch of the source never actually change, but the sound in respect to the observer's position makes it seem like the sound has changed. This effect can be observed when an emergency siren passes by an observer on the road. The siren sounds much higher in pitch as it approaches the observer and then lower after it passes and is getting farther away.

The Doppler effect also applies to situations involving light waves. An observer in space would see light approaching as having shorter wavelengths than it actually does, causing it to appear blue. When the light wave is getting farther away, the light would look red due to the apparent elongation of the wavelength. This is called the **red-blue shift**.

A recent addition to the study of waves is the gravitational wave. Its existence has been proven and verified, yet the details surrounding its capabilities are still under inquiry. Further understanding of gravitational waves could help scientists understand the beginnings of the universe and how the existence of the solar system is possible. This understanding could also include the future exploration of the universe.

Light

The movement of light is described like the movement of waves. Light travels with a wave front and has an **amplitude** (a height measured from the neutral), a cycle or wavelength, a period, and energy. Light travels at approximately 3.00×10^8 m/s and is faster than anything created by humans.

Light is commonly referred to by its measured **wavelengths**, or the length for it to complete one cycle. Types of light with the longest wavelengths include radio, TV, micro, and infrared waves. The next set of wavelengths are detectable by the human eye and make up the visible spectrum. The **visible spectrum** has wavelengths of 10^{-7} m, and the colors seen are red, orange, yellow, green, blue, indigo, and violet. Beyond the visible spectrum are even shorter wavelengths (also called the **electromagnetic spectrum**) containing ultraviolet light, x-rays, and gamma rays. The wavelengths outside of the visible light range can be harmful to humans if they are directly exposed, especially for long periods of time.

When a wave crosses a boundary or travels from one medium to another, certain actions take place. If the wave travels through one medium into another, it experiences **refraction,** which is the bending of the wave from one medium's density to another, altering the speed of the wave.

For example, a side view of a pencil in half a glass of water appears as though it is bent at the water level. What the viewer is seeing is the refraction of light waves traveling from the air into the water. Since the wave speed is slowed in water, the change makes the pencil appear bent.

When a wave hits a medium that it cannot pass through, it is bounced back in an action called **reflection.** For example, when light waves hit a mirror, they are reflected, or bounced off, the back of the mirror. This can cause it to seem like there is more light in the room due to the doubling back of the initial wave. This is also how people can see their reflection in a mirror.

When a wave travels through a slit or around an obstacle, it is known as **diffraction.** A light wave will bend around an obstacle or through a slit and cause a diffraction pattern. When the waves bend around an obstacle, additional waves and the spreading of light on the other side of the opening occurs.

Magnetism and Electricity

Magnetic forces occur naturally in specific types of materials and can be imparted to other types of materials. If two straight iron rods are observed, they will naturally have a negative end (pole) and a positive end (pole). These charged poles follow the rules of any charged item: Opposite charges attract, and like charges repel. When arranged positive to negative, they will attract each other, but if one rod is turned around, the two rods will repel each other due to the alignment of negative to negative poles and positive to positive poles. When poles are identified, magnetic fields are observed between them. If small iron filings (a material with natural magnetic properties) are sprinkled over a sheet of paper resting on top of a bar magnet, the field lines from the poles can be seen in the alignment of the iron filings, as pictured below:

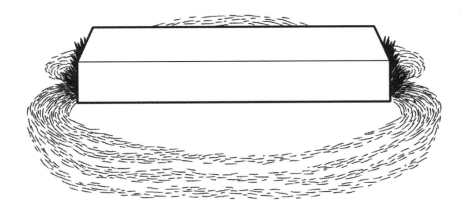

These fields naturally occur in materials with magnetic properties. There is a distinct pole at each end of such a material. If materials are not shaped with definitive ends, the fields will still be observed through the alignment of poles in the material. For example, a circular magnet does not have ends but still has a magnetic field associated with its shape, as pictured below:

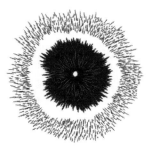

Magnetic forces can also be generated and amplified by using an electric current. For example, if an electric current is sent through a length of wire, it creates an electromagnetic field around the wire from the charge of the current. This force is from the moving of negatively-charged electrons from one end of the wire to the other. This is maintained as long as the flow of electricity is sustained. The magnetic field can also be used to attract and repel other items with magnetic properties. A smaller or larger magnetic force can be generated around this wire, depending on the strength of the current in the wire. As soon as the current is stopped, the magnetic force also stops.

Magnetic energy can be harnessed, or manipulated, from natural sources or from a generated source (a wire carrying electric current). When a core with magnetic properties (such as iron) has a wire wrapped around it in circular coils, it can be used to create a strong, non-permanent electromagnet. If current is run through the wrapped wire, it generates a magnetic field by polarizing the ends of the metal core, as described above, by moving the negative charge from one end to the other. If the direction of the current is reversed, so is the direction of the magnetic field due to the poles of the core being reversed. The term **non-permanent** refers to the fact that the magnetic field is generated only when the current is present, but not when the current is stopped. The following is a picture of a small electromagnet made from an iron nail, a wire, and a battery:

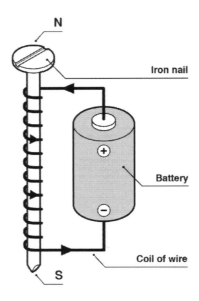

This type of electromagnetic field can be generated on a larger scale using more sizable components. This type of device is useful in the way it can be controlled. Rather than having to attempt to block a permanent magnetic field, the current to the system can simply be stopped, thus stopping the magnetic field. This provides the basis for many computer-related instruments and magnetic resonance imaging (MRI) technology. Magnetic forces are used in many modern applications, including the creation of super-speed transportation. Super magnets are used in rail systems and supply a cleaner form of energy than coal or gasoline. Another example of the use of super-magnets is seen in medical equipment, specifically MRI. These machines are highly sophisticated and useful in imaging the internal workings of the human body. For super-magnets to be useful, they often must be cooled down to extremely low temperatures to dissipate the amount of heat generated from their extended usage. This can be done by flooding the magnet with a super-cooled gas such as helium or liquid nitrogen. Much research is continuously done in this field to find new ceramic–metallic hybrid materials that have structures that can maintain their charge and temperature within specific guidelines for extended use.

Practice Questions

Example 1

Capacitors can be used for storing energy. Capacitors quickly discharge energy, unlike batteries. Among their applications, capacitors can be used in devices that need a sudden amount of energy at once, such as flash bulbs and defibrillators. When charged particles pass through a capacitor, it can change the direction of the charged particle. This technology is used in printers and televisions.

Most capacitors are made up of two parallel plates placed a slight distance apart. A complete circuit is made by connecting the positive end of a battery to one of the plates, and a negative end of the battery to the other plate.

Several capacitors are attached in a circuit with a 12 V battery, in an attempt to determine factors that could affect the capacitance, charge, and energy of each. The capacitors are different sizes, the plates are different distances (measured in mm) apart, and the plate sizes are different (measured in mm^2). Results are seen below:

Capacitor	Distance between plates (mm)	Area of plates (cm^2)	Capacitance (pF)	Charge (nC)
1	0.5	10	178	2.11
2	0.5	20	710	8.50
3	1.0	10	89.5	1.10
4	1.0	20	355	4.22
5	1.5	10	58	0.72
6	1.5	20	234	2.84

Capacitor 3 from the above chart is used to test the effects of dielectrics on capacitors. A dielectric is placed in between the two plates of a capacitor and multiplies the overall capacitance. The stronger the dielectric, the larger the increase in capacitance. Different materials were used as a dielectric and the following measurements for capacitance and charge were recorded in the chart below:

Dielectric material	Capacitance (pF)	Charge (nC)
Air (no dielectric)	89.5	1.10
Glass	687.0	8.21
Paper	266.5	3.20
Polystyrene	229.0	2.70
Quartz	335.4	4.04
Teflon	188.8	2.25

1. From the information in Chart 1, what type of capacitor would have the highest capacitance?
    a. Large distance between plates and large area
    b. Large distance between plates and small area
    c. Small distance between plates and large area
    d. Small distance between plates and small area

2. If another capacitor were added to Chart 1, with twice the distance between the plates and twice the area of Capacitor 6, what would be the charge on this new capacitor?
    a. 0.37 nC
    b. 2.66 nC
    c. 5.68 nC
    d. 11.31 nC

3. Which of the following correctly ranks the dielectrics in order of decreasing strength?
    a. Polystyrene, paper, quartz, glass
    b. Polystyrene, quartz, paper, glass
    c. Glass, paper, quartz, Teflon
    d. Glass, quartz, paper, Teflon

4. If another capacitor with plates of area 10 cm^2 and a distance between the plates of 1.25 mm were tested, which would be closest to the most likely resulting capacitance?
    a. 31 pF
    b. 58 pF
    c. 73.8 pF
    d. 89.5 pF

5. Which combination would result in the highest charge?
    a. Capacitor 1 with a paper dielectric
    b. Capacitor 2 with a glass dielectric
    c. Capacitor 3 with a Teflon dielectric
    d. Capacitor 4 with a quartz dielectric

Example 2

There are four stages in cell division: gap 1 (G1), synthesis (S), gap 2 (G2), and mitosis (M). The cell prepares for division in interphase during the G1, S, and G2 phases. Mitosis is when two cells form from the division of one cell.

The graph below shows the percentage of time a typical cell spends in each of these phases:

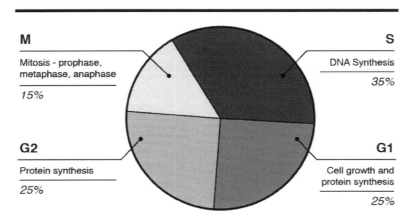

260

The figure below shows cyclin levels during cell cycle. Cyclins are a class of proteins that control cell division in mammalian cells:

Cell Cycle Stage

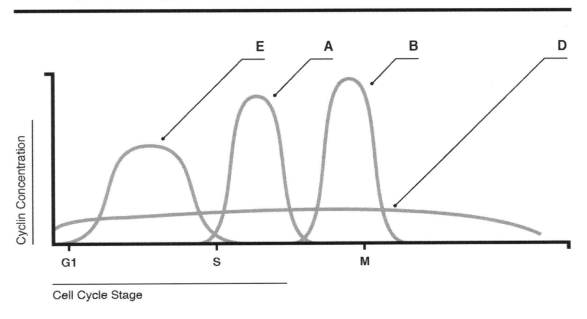

Cell Cycle Stage

6. A fibroblast (a mammalian cell) takes twenty-four hours to divide. How many hours would a fibroblast spend in mitosis?
    a. 12.0
    b. 8.4
    c. 6.0
    d. 3.6

7. Which of the following could be asserted regarding a mammalian cell beginning mitosis?
    a. Cyclin A must be increasing
    b. Cyclin B must be decreasing
    c. Cyclin A must be at its maximum level
    d. Cyclin E must be at its maximum level

8. Which cyclins MOST likely initiate the cell cycle?
    a. Only Cyclin A
    b. Only Cyclin B
    c. Only Cyclin E
    d. Only Cyclin A and Cyclin B

9. Which cyclin seems to have the LEAST effect on any phase of the mammalian cell cycle?
    a. Cyclin A
    b. Cyclin B
    c. Cyclin D
    d. Cyclin E

261

10. An increase in Cyclin A triggers which of the cell phases?
   a. Protein synthesis
   b. Mitosis
   c. DNA synthesis
   d. Cell growth

Example 3

The annual wildland fire statistics are compiled and reported by the National Interagency Coordination Center at NIFC for federal and state agencies. Information is gathered through situation reports from many decades.

The table below lists the number of fires during the years since 2007 and the acres those fires covered:

Total Wildland Fires and Acres (1926 - 2017)

The National Interagency Coordination Center at NIFC compiles annual wildland fire statistics for federal and state agencies. This information is provided through Situation Reports, which have been in use for several decades. Prior to 1983, sources of these figures are not known, or cannot be confirmed, and were not derived from the current situation reporting process. As a result, the figures prior to 1983 should not be compared to later data.

Source: National Interagency Coordination Center

Year	Fires	Acres
2017	71,499	10,026,086
2016	67,743	5,509,995
2015	68,151	10,125,149
2014	63,312	3,595,613
2013	47,579	4,319,546
2012	67,774	9,326,238
2011	74,126	8,711,367
2010	71,971	3,422,724
2009	78,792	5,921,786
2008	78,979	5,292,468
2007	85,795	9,328,045

The chart below lists temperature anomalies (both land and ocean) beginning in 1880 in degrees Celsius and degrees Fahrenheit:

Global Land and Ocean Temperature Anomalies, March

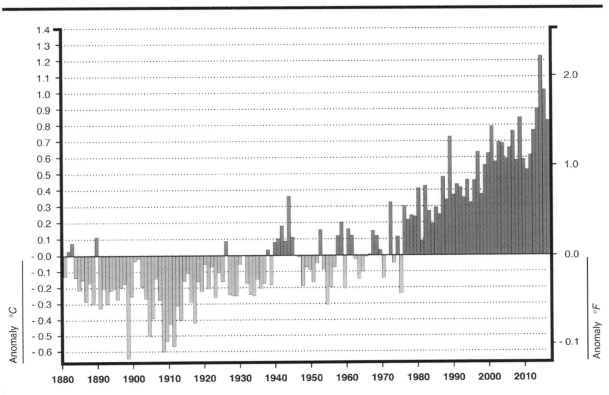

11. Approximately which year shows the highest temperature anomaly?
    a. 1900
    b. 1944
    c. 1990
    d. 2016

12. What was the average number of fires from 2007 to 2012?
    a. 67,774
    b. 76,240
    c. 70,512
    d. 78,792

13. What could be inferred about the years with the highest number of acres covered by its fires?
    a. There was a misreporting of the number of fires.
    b. The most acres reflected the greatest number of fires.
    c. The most acres reflected larger-sized fires.
    d. No conclusion can be drawn.

14. What could be inferred about the trend in temperature anomalies?
    a. Temperatures are increasing.
    b. Temperatures are decreasing.
    c. Temperatures have remained steady.
    d. Nothing can be inferred about the temperature anomalies.

15. Overall, what could NOT have caused the number of total fires to decrease?
    a. Better policing of camp grounds
    b. Improved fire-fighting technologies
    c. Increased fire-fighting awareness
    d. Decrease in climate temperatures

Example 4

When the materials of surfaces move across one another they have a specific coefficient of friction that impedes their movement. In order to begin the sliding motion of these surfaces, a coefficient of static friction must be overcome. Once the materials begin moving across each other, the coefficient of kinetic friction is less than that of what it took to begin the sliding motion. The coefficient of friction does not have units, as it represents a proportion.

The table below lists the coefficients of static and kinetic friction of rubber on various materials:

Material 1	Material 2	Kinetic friction	Static friction
Rubber	Concrete	0.67	0.89
Rubber	Asphalt	0.66	0.84
Rubber	Cardboard	0.54	0.79
Rubber	Ice	0.14	

The graph below shows the relationship between the coefficient of friction and the distance moved for rubber on three of the materials:

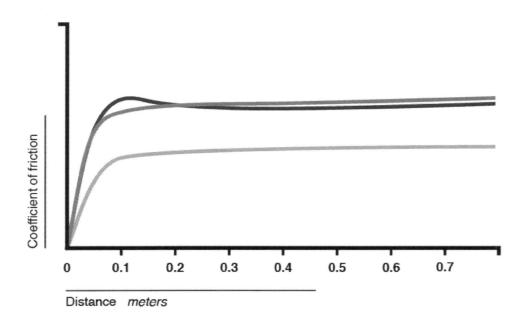

Distance *meters*

264

16. What is the average coefficient of kinetic friction for the materials listed?
    a. 0.67
    b. 0.50
    c. 0.40
    d. 0.84

17. If material X is slightly rougher than cardboard, but smoother than asphalt, which of the following could be an approximation for the coefficient of kinetic friction of rubber on material X?
    a. 0.54
    b. 0.60
    c. 0.66
    d. Cannot be determined

18. Which of the following would describe the graphs of the coefficient of friction versus distance for the given materials?
    a. Linear
    b. Quadratic
    c. Exponential decay
    d. Logarithmic

19. Which color line in the graph would most likely match up to the information for concrete?
    a. Black
    b. Dark gray (the line that is higher up on the y-axis)
    c. Light gray (the line that is lower down on the y-axis)
    d. None would match concrete

20. Why do the lines in Graph 1 all seem to level off just before a certain reading on the coefficient of friction?
    a. Miscalculations
    b. They are all nearly the same material
    c. The sliding friction is overcome at that point
    d. A lubricant is added at that point

Example 5

Our solar system is made up of eight planets. In order from the Sun moving outward, they are Mercury, Venus, Earth, Mars, Jupiter, Saturn, Uranus, and Neptune. The first four planets are the smallest and have some similarities in characteristics, while the outer four planets are larger and also have some comparable traits. Kepler studied the movements of the planets and devised three laws for planetary motion.

Some basic information regarding the planets is listed in the tables below:

Planet Name	Distance from Sun (km)	Mass (10^{24}kg)	Gravity (m/s^2)
Mercury	57.9 million	0.330	3.7
Venus	108.2 million	4.87	8.9
Earth	149.6 million	5.97	9.8
Mars	227.9 million	0.642	3.7
Jupiter	778.6 million	1898	23.1
Saturn	1,433.5 million	568	9.0
Uranus	2,872.5 million	86.8	8.7
Neptune	4,495.1 million	102	11.0

Planet	Mercury	Venus	Earth	Mars	Jupiter	Saturn	Uranus	Neptune
Length of Orbit (Days)	88.0	224.7	365.2	687	4,331	10,747	30,589	59,800
Number of Moons	0	0	1	2	67	62	27	14

21. The mass of Venus is approximately what percentage of the Earth's mass?
    a. 8.16 percent
    b. 81.6 percent
    c. 1.22 percent
    d. 12.2 percent

22. What would NOT be a likely reason for the variations in gravity on each planet?
    a. The pull of the moons on a planet
    b. The speed of rotation of a planet
    c. The make-up of the core of a planet
    d. The mass of a planet

23. What is the average length of orbit, in days, of the planets listed?
    a. 365.2
    b. 13,354
    c. 29,944
    d. 59,800

24. How many times farther is Neptune from the Sun than Mercury, and how many times longer is Neptune's orbit than Earth's?
 a. 78 times as far, and 164 times as long
 b. 78 times as far, and 680 times as long
 c. 30 times as far, and 680 times as long
 d. 30 times as far, and 164 times as long

25. Which distance is the shortest?
 a. Distance from Mercury to the Sun
 b. Distance from Mercury to Venus
 c. Distance from Venus to Earth
 d. Distance from Mars to Earth

Example 6

Soil samples were collected from various locations and analyzed for their composition. These minerals were sand, silt, and clay. Three types of minerals were identified and measured by percent in each sample. Particle size ranges for each mineral were also measured and recorded for the soil samples.

Soil Sample	Sand (%)	Clay (%)	Silt (%)
1	75	5	20
2	5	80	15
3	20	35	45
4	70	15	15
5	55	25	20

Type of mineral particle	Size range of particles (mm)
Sand	3.0-0.07
Silt	0.07-0.003
Clay	Less than 0.003

26. Which minerals mainly comprised Sample 3?
 a. Sand and silt
 b. Silt and clay
 c. Sand and clay
 d. Sand

27. Which soil sample would MOST likely have particle sizes around 1.7 mm?
 a. Sample 2
 b. Sample 3
 c. Sample 4
 d. Sample 5

28. If the particles in Sample 2 were measured, what would best match their average size?
    a. 0.001 mm
    b. 0.06 mm
    c. 1.0 mm
    d. 1.5 mm

29. If another sample was collected from an area near that of Sample 4, what size could the particles be from this new sample?
    a. Larger than 3.0 mm
    b. Larger than 0.07 mm, but smaller than 3.0 mm
    c. Smaller than 0.07 mm, but larger than 0.003 mm
    d. Smaller than 0.003 mm

30. Which would describe the make-up of Sample 1?
    a. More clay than sand
    b. More silt than sand
    c. Less sand than clay
    d. Less clay than sand

Example 7

Dendrochronology is a method of dating using tree growth through the counting of concentric bands in the trunk. This is not a completely accurate method of calculation; therefore, it is often paired with other methods of cross dating, involving referencing characteristics of rings with other samples in an area of similar conditions. Variations in bands can be caused by environmental conditions such as annual rainfall, as the rings were formed. When there is less rainfall, fewer rings are noticed to be formed. Rings that are formed during this time are often narrower than in times of heavier rainfall, as heavier rain can result in faster growth.

Scientists studied oak trees at three separate sites and compiled the following data. The number of trees sampled was greater than 50.

Site	Average number of growth bands per year	Average size of growth bands (mm)
1	11	2
2	14	3
3	19	11

31. Which site received the most rainfall?
    a. Site 1
    b. Site 2
    c. Site 3
    d. Cannot be determined

32. From the graph below, which would best represent the relationship between the average number of growth bands verses the average size of growth bands?

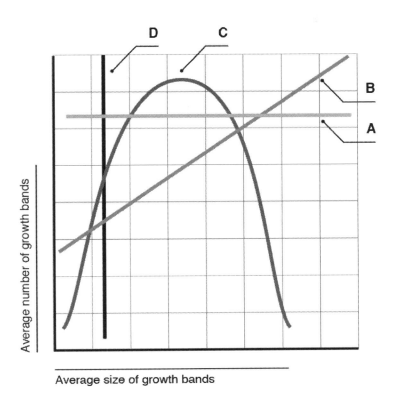

a. A
b. B
c. C
d. D

33. Which of the following would describe the trees at Site 1 when compared to those at Site 2?
    a. The trees at Site 1 were not homogeneous, but the trees at Site 2 were.
    b. The trees at Site 1 experienced the same growth rate as the trees at Site 2.
    c. The trees at Site 1 experienced a slower growth rate than the trees at Site 2.
    d. The trees at Site 1 experienced a faster growth rate than the trees at Site 2.

34. Why is cross dating done?
    a. To decrease the number of trees necessary for study
    b. To predict the amount of inches of rainfall for a specific area
    c. To improve the accuracy when predicting growth rate for trees
    d. To decrease the number of bands used for calculating rainfall for a specific area

35. Trees from another site were measured and found to have an average of 17 growth bands per year. When compared to the data in Table 1, what would likely be the average size of these growth bands?

  a. Less than 2 mm
  b. Between 2 mm and 4 mm
  c. Between 4 mm and 12 mm
  d. Greater than 12 mm

Example 8

Atoms of a chemical element, or radioactive isotope, naturally go through a process of decay called radioactive decay. Through this process, the atom decays into another element by emitting alpha particles, beta particles, and gamma rays. This process alters the composition of the atom's nucleus until it reaches a stable state. The amount of time it takes for half (50 percent) of the atoms in a sample to decay is referred to as half-life. The figure below shows the decay of Radon 222 into Polonium 218 and other decay products:

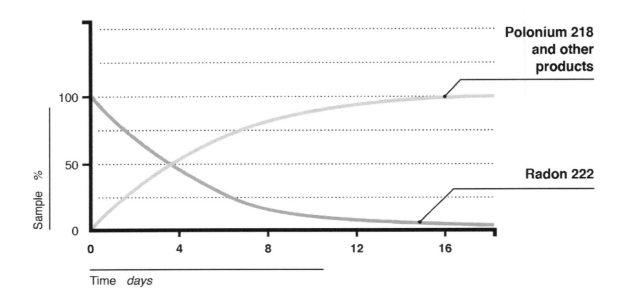

The figure below shows the decay from Mercury 206 into Thallium 206 into Lead 206:

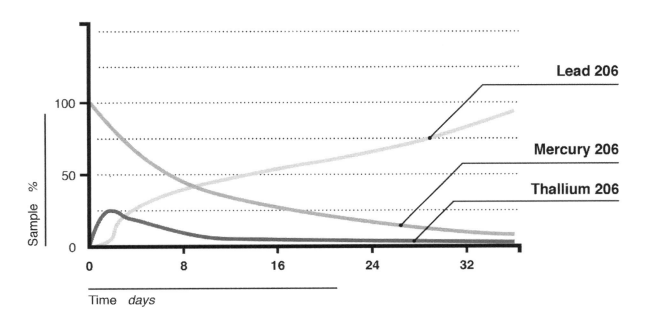

The table below displays the decay products and associated energy million electron volts (MeV) and the type of particle emitted with the decay:

Isotope Particle	Decay Product	Energy (MeV)	Decay Type
Radon 222	Polonium 218	6.190	Alpha
Lead 210	Mercury 206	2.992	Alpha
Mercury 206	Thallium 206	0.912	Beta
Thallium 206	Lead 206	0.813	Beta

36. What is the approximate half-life of Radon 222?
    a. 16 days
    b. 12 days
    c. 4 days
    d. 2 days

37. Which statement is true?
    a. Radioactive decay products are all unstable.
    b. Radioactive decay never occurs naturally.
    c. Radioactive decay is a natural process.
    d. Radioactive decay occurs in half of a sample.

38. Which relationship can be determined from the decay energy and the type of decay particle?
    a. Beta particles have higher decay energies.
    b. Alpha and beta particles have similar decay energies.
    c. Alpha particles have higher decay energies.
    d. There is no relationship between particles and decay energies.

39. When do Radon 222 and Polonium 218 have approximately the same number of atoms remaining?
    a. Day 16
    b. Day 8
    c. Day 4
    d. Day 2

40. Which of the following best describes the Mercury 206 curve in Figure 2?
    a. The rate of decay is a linear function because the number of atoms converting to other elements is constant.
    b. The rate of decay occurs quickly at first, then slows as the number of atoms is reduced.
    c. The rate of decay occurs slowly at first, then speeds up as the number of atoms converting increases.
    d. The rate of decay is sinusoidal because the number of atoms converting to other elements constantly goes up and down.

Answer Explanations

1. C: Looking at Table 1, a comparison of Capacitor 1 to Capacitor 3 shows that the area between the plates is the same, but the distance between the plates varies. The distance between the plates of Capacitor 1 is smaller than the distance between the plates of Capacitor 3, and Capacitor 1 has a greater capacitance. A comparison of Capacitor 2 to Capacitor 1 shows that the distance between the plates is the same, but the area of the plates varies. The area of Capacitor 2's plates is larger with a larger capacitance. This shows that a smaller distance and a larger area will result in a larger capacitance.

2. C: Looking at Table 1, a comparison of the capacitance of Capacitors 1 and 4 shows the relationship of doubling the area and the distance, which results in a doubling of the charge. Therefore, if another capacitor were added with twice the area and twice the distance of Capacitor 6, it would double the charge, which would be 5.68 nC.

3. D: The information states that a stronger dielectric would result in a larger capacitance. Looking at Chart 2 from the question, the dielectrics in order from largest capacitance to smallest capacitance are glass, quartz, paper, and Teflon.

4. C: Using Table 1, Capacitors 3 and 5 both have the same area for their plates. Capacitor 3 has a distance of 1.0 mm and a capacitance of 89.5 pF, while Capacitor 5 has a distance of 1.5 mm and a capacitance of 58 pF. Therefore, a capacitor with a distance between its plates halfway between Capacitor 3 and Capacitor 5 would have a capacitance approximately halfway between their capacitances, which would be 73.8 pF.

5. B: Combining the charges from the listed capacitors in Table 1 with the charge from the respective dielectrics in Table 2 results in the largest combination being that of Capacitor 2 (at 8.50 nC) and glass (at 8.21 nC).

6. D: Looking at Figure 1 shows that mitosis takes up 15 percent of the entire cycle, and 15 percent of twenty-four hours is 3.6 hours.

7. B: Looking at Figure 2 shows that Cyclin B is decreasing during the mitosis (M) stage of the cell cycle.

8. C: Looking at Figure 2, Cyclin E is the only choice that is present at the beginning of the cell cycle.

9. C: Looking at Figure 2, Cyclin D is the only choice that seems to be steady throughout the entire cell cycle; thus, Cyclin D will most likely have the least effect on the cycle.

10. C: Looking at Figure 2, the increase in Cyclin A happens at the S phase of the cell cycle. Referencing the cycle components in Figure 1 shows that DNA synthesis is the only occurrence in Phase S.

11. D: For the years listed on Chart 2, Choice *D* shows an anomaly of approximately 2.22 degrees Fahrenheit, Choice *A* shows an anomaly of approximately -1.22 degrees Fahrenheit, Choice *B* shows an anomaly of approximately 0.7 degrees Fahrenheit, and Choice *C* shows an anomaly of approximately 1.46 degrees Fahrenheit. Overall, Choice *D* is the highest.

12. B: Looking at Table 1, the number listed for 2012 is Choice *A*, 67,774. The number listed for 2008 is Choice *D*, 78,792. Neither of these represents the average. In order to calculate the average, the following formula should be used:

$$\frac{\Sigma \text{ fires}}{Number\ of\ values} = \frac{457,437}{6} = 76,240$$

Choice *B* lists the correct value for the average number of fires and Choice *C* is the average for 2007 to 2017, which is erroneous because the question only requests the average of number of fires from 2007 to 2012.

13. C: Because the highest number of acres that are listed are not paired with the largest numbers of fires, another factor would have to contribute to the size of the fires. Many of the larger numbers in acres are after a year with a higher temperature anomaly. This potentially indicates possible drought conditions, which would make it more difficult to contain or extinguish even a small number of fires.

14. A: Chart 2 shows an increase in temperatures, thus causing an increase in anomalies. There have not been any decreases in temperatures significant enough to cause an anomaly since the 1970s; thus, Choice *B* is not correct. The chart displays multiple anomalies, so Choices *C* and *D* are also not correct.

15. D: There is no way to confirm the assertions of Choices *A*, *B*, or *C*, but all would contribute to the reduction of accidental fires and fires in general, while Choice *D* is not displayed through the information represented in Chart 2. Therefore, decreased climate temperatures are the only verifiable thing not contributing to the overall decrease in the number of fires.

16. B: Choice *B* is calculated using the following:

$$\frac{\Sigma \text{ coefficients}}{Number\ of\ values} = \frac{2.01}{4} = 0.50$$

Choice *A* is an extreme value listed, not an average. Choice *D* is the average of static friction (not kinetic friction). Choice *C* is a miscalculation of the average because it takes the sum of the coefficients and erroneously divides it by five, even though there are only four values contributing to the average.

17. B: Choice *B* gives a value that is partway between the coefficient of kinetic friction of cardboard and the coefficient of kinetic friction of asphalt, which would be a reasonable approximation for material X. Choice *A* would be too low, representing the coefficient of kinetic friction of cardboard, and Choice *C* is too high, representing the coefficient of kinetic friction of asphalt. Choice *D* is not correct, because a relationship can be established between the materials.

18. D: The shape of the graph is most like that of a logarithmic function; therefore, Choice *D* is the only possible correct answer. Choice *A*, linear, would look like:

Linear function

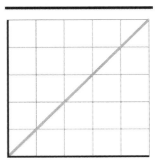

Choice *B*, quadratic, would look like:

Quadratic function

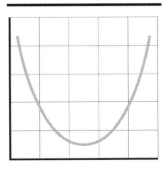

Choice *C*, exponential decay, would look like:

Exponential Decay

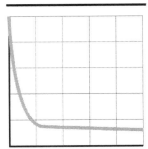

19. A: Looking at Table 1, rubber and concrete have the highest coefficient of static friction, and therefore would match the graph that matches the highest value on the *y*-axis (vertical). None of the other choices model this extreme.

20. C: The value of static friction for surfaces in contact must be overcome in order to start motion. After this coefficient of static friction is overcome, the materials begin moving across each other, and the coefficient of kinetic friction is less than that of what it took to begin the sliding motion. As two out of the three graphed lines are similar in value as to where the coefficient of static friction is overcome, it shows that the surfaces move more easily across each other with the lower coefficient of kinetic friction.

21. B: Choice *B* correctly calculates the relative percentage that the mass of Venus is relative to that of Earth mass by utilizing the following:

$$mass\ of\ Venus = x\% \times Earth's\ mass$$

$$4.87 = x\% \times 5.97$$

$$x = 81.6\%$$

Choice *A* does not move the decimal correctly to convert to a percent; Choice *C* reverses the values for the mass of Venus and the mass of Earth in the calculations; and Choice *D* reverses the values for the mass of Venus and the mass of Earth in the calculations, and it also does not move the decimal correctly to convert to a percent.

22. A: Not all planets in the solar system have moons, so Choice *A* would not be a cause for variation in the gravity of a planet. Choices *B, C,* and *D* all have to do with characteristics that could potentially influence a gravitational pull on a planet.

23. B: Choice *B* is the only option that correctly calculates the average orbit, in days, using the following:

$$\frac{\Sigma\ orbits}{number\ of\ values} = \frac{106{,}831.9}{8} = 13{,}354$$

Choice *A* is the orbit of Earth, Choice *C* is an incorrect calculation of the average, and Choice *D* is the orbit of Neptune.

24. A: Choice *A* is the only correct combination. To calculate the distance of Neptune from the Sun relative to that of Mercury to the Sun, the following should be used:

$$\frac{4495.1}{57.9} = 78\ times\ as\ far$$

In order to calculate the orbit time of Neptune relative to that of Earth, the following should be used:

$$\frac{59800}{365.2} = 164\ times\ as\ long$$

Choices *B, C,* and *D* are incorrect because they use the relative orbit time of Mercury and Neptune, the relative distance from the sun of Earth and Neptune, or both.

25. C: Choice *C* is the closest at 41.4 million km apart, while Choice *B* is next at 50.3 million km apart. Next is Choice *A* at 57.9 million km apart, and the furthest is Choice *D* at 78.3 million km apart.

26. B: According to Table 1, Sample 3 is made up of 45 percent silt and 35 percent clay. Sand comprised only 20 percent of the sample. Therefore Choices *A, C,* and *D* are incorrect.

27. C: According to Table 2, sand particles can have a size of 1.7 mm. As sample 4 is the only sample listed from Table 1 that has mostly sand as its percent composition, Choice *C* is the best option. Choice *D*, Sample 5, also has a higher percent of sand, but still less than Sample 4.

28. A: According to Table 1, Sample 2 is primarily composed of clay at 80 percent. Looking at Table 2, clay particles range from 0–0.003 mm, which would give an average of 0.0015 mm. All of the other choices are larger than even the largest possible range for clay, so they could not be answers for Sample 2.

29. B: If the new sample were similar in composition to Sample 4, it would have mostly sand particles. From Table 2, sand particle sizes fall in the range of 0.07–3.0 mm; therefore, the best choice would be for the particles to be larger than the minimum size for sand. Choice *A* is larger than any of the particle options listed, and Choices *C* and *D* are too small for the size range of sand.

30. D: According to Table 1, Sample 1 contains 75 percent sand, 5 percent clay, and 20 percent silt. This can be expressed by sand > silt > clay. Therefore, only Choice *D* is correct.

31. C: According to the information provided, more rainfall produced more growth bands per year and an increased size of the growth bands. Site 3 had both the greatest number of growth bands and the largest average size of growth bands; therefore, Choice *C* is the best possible answer.

32. B: The graph for Choice *B* is the only one that displays a linear relationship, which shows that as the number of growth bands increases, so does the size of the growth bands. Graphs for Choices *A* and *D* show that one of the factors is held constant (which is not the case for any of the sites) and that for Choice *C* shows an increase in both factors and then a decrease in the number of growth bands.

33. C: Site 1 had smaller growth bands in both amount and size than those of Site 2. This leads to the conclusion that the trees at Site 1 did not grow as fast as the trees at Site 2.

34. C: The information provided relays that counting the number of growth bands is not completely accurate on its own. Therefore, the cross dating can help to increase the accuracy for predicting the growth rate of trees. Decreasing the number of trees studied or the number of bands studied could decrease the accuracy of a study. There is no way to predict the actual amount of rainfall associated with the band number or size according to this study; therefore, Choice *C* is the best answer.

35. C: According to the data in Table 1, the number of growth bands at the new site falls between the number found at Site 2 and the number found at Site 3. Therefore, the average size of these growth bands should be somewhere between the sizes for Site 1 and Site 3, which is between 4 mm and 12 mm. The other choices are too small or too big for a comparison between Site 1 and Site 2.

36. C: According to Figure 1, 50 percent of the sample remains after approximately four days. This is where the graph lines up with 50 percent on the y-axis and drops down to four days on the x-axis.

37. C: As stated in the example, radioactive decay is a natural process that happens spontaneously when atoms of one element decay into another element. Choice *A* is incorrect because decay occurs until the atoms in the sample reach a stable state. It is not limited to only half of the atoms; therefore, Choice *D* is incorrect.

38. C: According to Table 1, the higher energy emissions were matched with the alpha particle emissions. The alpha particles had higher emissions than the beta particles in all instances listed; there was a relationship between the emission energies and the types of particles emitted, so Choice *D* is incorrect.

39. C: According to Figure 1, the two lines intersect with each other at approximately Day 4.

40. B: The rate of decay occurs quickly at first, then slows as the number of Mercury 206 atoms is reduced.

ACT Writing Test

Elements of the Writing Process

Skilled writers undergo a series of steps that comprise the writing process. The purpose of adhering to a structured approach to writing is to develop clear, meaningful, coherent work.

The stages are pre-writing or planning, organizing, drafting/writing, revising, and editing. Not every writer will necessarily follow all five stages for every project, but will judiciously employ the crucial components of the stages for most formal or important work. For example, a brief informal response to a short reading passage may not necessitate the need for significant organization after idea generation, but larger assignments and essays will likely mandate use of the full process.

Pre-Writing/Planning
Brainstorming
One of the most important steps in writing is pre-writing. Before drafting an essay or other assignment, it's helpful to think about the topic for a moment or two, in order to gain a more solid understanding of what the task is. Then, spend about five minutes jotting down the immediate ideas that could work for the essay. **Brainstorming** is a way to get some words on the page and offer a reference for ideas when drafting. Scratch paper is provided for writers to use any pre-writing techniques such as webbing, freewriting, or listing. Some writers prefer using graphic organizers during this phase. The goal is to get ideas out of the mind and onto the page.

Freewriting
Like brainstorming, **freewriting** is another prewriting activity to help the writer generate ideas. This method involves setting a timer for two or three minutes and writing down all ideas that come to mind about the topic using complete sentences. Once time is up, writers should review the sentences to see what observations have been made and how these ideas might translate into a more unified direction for the topic. Even if sentences lack sense as a whole, freewriting is an excellent way to get ideas onto the page in the very beginning stages of writing. Using complete sentences can make this a bit more challenging than brainstorming, but overall it is a worthwhile exercise, as it may force the writer to come up with more complete thoughts about the topic.

Once the ideas are on the page, it's time for the writer to turn them into a solid plan for the essay. The best ideas from the brainstorming results can then be developed into a more formal outline.

Organizing
Although sometimes it is difficult to get going on the brainstorming or prewriting phase, once ideas start flowing, writers often find that they have amassed too many thoughts that will not make for a cohesive and unified essay. During the organization stage, writers should examine the generated ideas, hone in on the important ones central to their main idea, and arrange the points in a logical and effective manner. Writers may also determine that some of the ideas generated in the planning process need further elaboration, potentially necessitating the need for research to gather infortmation to fill the gaps.

Once a writer has chosen his or her thesis and main argument, selected the most applicable details and evidence, and eliminated the "clutter," it is time to strategically organize the ideas. This is often accomplished with an outline.

Outlining

An **outline** is a system used to organize writing. When composing essays, outlining is important because it helps writers organize important information in a logical pattern using Roman numerals. Usually, outlines start out with the main ideas and then branch out into subgroups or subsidiary thoughts or subjects. Not only do outlines provide a visual tool for writers to reflect on how events, ideas, evidence, or other key parts of the argument relate to one another, but they can also lead writers to a stronger conclusion. The sample below demonstrates what a general outline looks like:

I. Introduction
    1. Background
    2. Thesis statement
II. Body
    1. Point A
        a. Supporting evidence
        b. Supporting evidence
    2. Point B
        a. Supporting evidence
        b. Supporting evidence
    3. Point C
        a. Supporting evidence
        b. Supporting evidence
III. Conclusion
    1. Restate main points of the paper.
    2. End with something memorable.

Drafting/Writing

Now it comes time to actually write the essay. In this stage, writers should follow the outline they developed in the brainstorming process and try to incorporate the useful sentences penned in the freewriting exercise. The main goal of this phase is to put all the thoughts together in cohesive sentences and paragraphs.

It is helpful for writers to remember that their work here does not have to be perfect. This process is often referred to as **drafting** because writers are just creating a rough draft of their work. Because of this, writers should avoid getting bogged down on the small details.

Referencing Sources

Anytime a writer quotes or paraphrases another text, they will need to include a citation. A **citation** is a short description of the work that a quote or information came from. The style manual your teacher wants you to follow will dictate exactly how to format that citation. For example, this is how one would cite a book according to the APA manual of style:

- *Format:* Last name, First initial, Middle initial. (Year Published) *Book Title.* City, State: Publisher.
- *Example:* Sampson, M. R. (1989). *Diaries from an Alien Invasion. Springfield, IL:* Campbell Press.

Revising

Revising offers an opportunity for writers to polish things up. Putting one's self in the reader's shoes and focusing on what the essay actually says helps writers identify problems—it's a movement from the mindset of writer to the mindset of editor. The goal is to have a clean, clear copy of the essay.

The main goal of the revision phase is to improve the essay's flow, cohesiveness, readability, and focus. For example, an essay will make a less persuasive argument if the various pieces of evidence are scattered and presented illogically or clouded with unnecessary thought. Therefore, writers should consider their essay's structure and organization, ensuring that there are smooth transitions between sentences and paragraphs. There should be a discernable introduction and conclusion as well, as these crucial components of an essay provide readers with a blueprint to follow.

Additionally, if the writer includes copious details that do little to enhance the argument, they may actually distract readers from focusing on the main ideas and detract from the strength of their work. The ultimate goal is to retain the purpose or focus of the essay and provide a reader-friendly experience. Because of this, writers often need to delete parts of their essay to improve its flow and focus. Removing sentences, entire paragraphs, or large chunks of writing can be one of the toughest parts of the writing process because it is difficult to part with work one has done. However, ultimately, these types of cuts can significantly improve one's essay.

Lastly, writers should consider their voice and word choice. The voice should be consistent throughout and maintain a balance between an authoritative and warm style, to both inform and engage readers. One way to alter voice is through word choice. Writers should consider changing weak verbs to stronger ones and selecting more precise language in areas where wording is vague. In some cases, it is useful to modify sentence beginnings or to combine or split up sentences to provide a more varied sentence structure.

Editing
Rather than focusing on content (as is the aim in the revising stage), the **editing** phase is all about the mechanics of the essay: the syntax, word choice, and grammar. This can be considered the proofreading stage. Successful editing is what sets apart a messy essay from a polished document.

The following areas should be considered when proofreading:

- Sentence fragments
- Awkward sentence structure
- Run-on sentences
- Incorrect word choice
- Grammatical agreement errors
- Spelling errors
- Punctuation errors
- Capitalization errors

One of the most effective ways of identifying grammatical errors, awkward phrases, or unclear sentences is to read the essay out loud. Listening to one's own work can help move the writer from simply the author to the reader.

During the editing phase, it's also important to ensure the essay follows the correct formatting and citation rules as dictated by the assignment.

Recursive Writing Process
While the writing process may have specific steps, the good news is that the process is recursive, meaning the steps need not be completed in a particular order. Many writers find that they complete steps at the same time such as drafting and revising, where the writing and rearranging of ideas occur simultaneously or in very close order. Similarly, a writer may find that a particular section of a draft needs more development, and will go back to the prewriting stage to generate new ideas. The steps can be repeated

at any time, and the more these steps of the recursive writing process are employed, the better the final product will be.

Practice Makes Prepared Writers

Like any other useful skill, writing only improves with practice. While writing may come more easily to some than others, it is still a skill to be honed and improved. Regardless of a person's natural abilities, there is always room for growth in writing. Practicing the basic skills of writing can aid in preparations for the ACT.

One way to build vocabulary and enhance exposure to the written word is through reading. This can be through reading books, but reading of any materials such as newspapers, magazines, and even social media count towards practice with the written word. This also helps to enhance critical reading and thinking skills, through analysis of the ideas and concepts read. Think of each new reading experience as a chance to sharpen these skills.

Writing Prompt

Directions

Write an organized, coherent essay about the idea of freedom of speech. In your essay, make sure you:

- State your own opinion on the topic and analyze the relationship between your opinion and at least one other opinion

- Develop and support your ideas with reasoning and examples

- Organize your ideas in a logical way

- Communicate your ideas efficiently in standard written English

Your opinion can be in full agreement with any of those given, in partial agreement, or completely different.

Sample Prompt

The true boundary line of the First Amendment can be fixed only when Congress and the courts realize that the principle on which speech is classified as lawful or unlawful involves the balancing against each other of two very important social interests, in public safety and in the search for truth. Every reasonable attempt should be made to maintain both interests unimpaired, and the great interest in free speech should be sacrificed only when the interest in public safety is really imperiled, and not, as most men believe, when it is barely conceivable that it may be slightly affected. In war time, therefore, speech should be unrestricted by the censorship or by punishment, unless it is clearly liable to cause direct and dangerous interference with the conduct of the war.

Greetings!

First, we would like to give a huge "thank you" for choosing us and this study guide for your ACT exam. We hope that it will lead you to success on this exam and for your years to come.

Our team has tried to make your preparations as thorough as possible by covering all of the topics you should be expected to know. In addition, our writers attempted to create practice questions identical to what you will see on the day of your actual test. We have also included many test-taking strategies to help you learn the material, maintain the knowledge, and take the test with confidence.

We strive for excellence in our products, and if you have any comments or concerns over the quality of something in this study guide, please send us an email so that we may improve.

As you continue forward in life, we would like to remain alongside you with other books and study guides in our library, such as;

ACCUPLACER: amazon.com/dp/1628457724

SAT: amazon.com/dp/1628457856

We are continually producing and updating study guides in several different subjects. If you are looking for something in particular, all of our products are available on Amazon. You may also send us an email!

Sincerely,
APEX Test Prep
info@apexprep.com

FREE

Free Study Tips DVD

In addition to the tips and content in this guide, we have created a FREE DVD with helpful study tips to further assist your exam preparation. **This FREE Study Tips DVD provides you with top-notch tips to conquer your exam and reach your goals.**

Our simple request in exchange for the strategy-packed DVD is that you email us your feedback about our study guide. We would love to hear what you thought about the guide, and we welcome any and all feedback—positive, negative, or neutral. It is our #1 goal to provide you with top quality products and customer service.

To receive your **FREE Study Tips DVD**, email freedvd@apexprep.com. Please put "FREE DVD" in the subject line and put the following in the email:

> a. The name of the study guide you purchased.
>
> b. Your rating of the study guide on a scale of 1-5, with 5 being the highest score.
>
> c. Any thoughts or feedback about your study guide.
>
> d. Your first and last name and your mailing address, so we know where to send your free DVD!

Thank you!

Made in United States
Orlando, FL
10 November 2021